EXPERIENCE IN THE MANAGEMENT OF RADIOACTIVE WASTE AFTER NUCLEAR ACCIDENTS: A BASIS FOR PREPLANNING

The following States are Members of the International Atomic Energy Agency:

AFGHANISTAN
ALBANIA
ALGERIA
ANGOLA
ANTIGUA AND BARBUDA
ARGENTINA
ARMENIA
AUSTRALIA
AUSTRIA
AZERBAIJAN
BAHAMAS
BAHRAIN
BANGLADESH
BARBADOS
BELARUS
BELGIUM
BELIZE
BENIN
BOLIVIA, PLURINATIONAL STATE OF
BOSNIA AND HERZEGOVINA
BOTSWANA
BRAZIL
BRUNEI DARUSSALAM
BULGARIA
BURKINA FASO
BURUNDI
CAMBODIA
CAMEROON
CANADA
CENTRAL AFRICAN REPUBLIC
CHAD
CHILE
CHINA
COLOMBIA
COMOROS
CONGO
COSTA RICA
CÔTE D'IVOIRE
CROATIA
CUBA
CYPRUS
CZECH REPUBLIC
DEMOCRATIC REPUBLIC OF THE CONGO
DENMARK
DJIBOUTI
DOMINICA
DOMINICAN REPUBLIC
ECUADOR
EGYPT
EL SALVADOR
ERITREA
ESTONIA
ESWATINI
ETHIOPIA
FIJI
FINLAND
FRANCE
GABON
GEORGIA
GERMANY
GHANA
GREECE
GRENADA
GUATEMALA
GUYANA
HAITI
HOLY SEE
HONDURAS
HUNGARY
ICELAND
INDIA
INDONESIA
IRAN, ISLAMIC REPUBLIC OF
IRAQ
IRELAND
ISRAEL
ITALY
JAMAICA
JAPAN
JORDAN
KAZAKHSTAN
KENYA
KOREA, REPUBLIC OF
KUWAIT
KYRGYZSTAN
LAO PEOPLE'S DEMOCRATIC REPUBLIC
LATVIA
LEBANON
LESOTHO
LIBERIA
LIBYA
LIECHTENSTEIN
LITHUANIA
LUXEMBOURG
MADAGASCAR
MALAWI
MALAYSIA
MALI
MALTA
MARSHALL ISLANDS
MAURITANIA
MAURITIUS
MEXICO
MONACO
MONGOLIA
MONTENEGRO
MOROCCO
MOZAMBIQUE
MYANMAR
NAMIBIA
NEPAL
NETHERLANDS
NEW ZEALAND
NICARAGUA
NIGER
NIGERIA
NORTH MACEDONIA
NORWAY
OMAN
PAKISTAN
PALAU
PANAMA
PAPUA NEW GUINEA
PARAGUAY
PERU
PHILIPPINES
POLAND
PORTUGAL
QATAR
REPUBLIC OF MOLDOVA
ROMANIA
RUSSIAN FEDERATION
RWANDA
SAINT KITTS AND NEVIS
SAINT LUCIA
SAINT VINCENT AND THE GRENADINES
SAMOA
SAN MARINO
SAUDI ARABIA
SENEGAL
SERBIA
SEYCHELLES
SIERRA LEONE
SINGAPORE
SLOVAKIA
SLOVENIA
SOUTH AFRICA
SPAIN
SRI LANKA
SUDAN
SWEDEN
SWITZERLAND
SYRIAN ARAB REPUBLIC
TAJIKISTAN
THAILAND
TOGO
TONGA
TRINIDAD AND TOBAGO
TUNISIA
TÜRKİYE
TURKMENISTAN
UGANDA
UKRAINE
UNITED ARAB EMIRATES
UNITED KINGDOM OF GREAT BRITAIN AND NORTHERN IRELAND
UNITED REPUBLIC OF TANZANIA
UNITED STATES OF AMERICA
URUGUAY
UZBEKISTAN
VANUATU
VENEZUELA, BOLIVARIAN REPUBLIC OF
VIET NAM
YEMEN
ZAMBIA
ZIMBABWE

The Agency's Statute was approved on 23 October 1956 by the Conference on the Statute of the IAEA held at United Nations Headquarters, New York; it entered into force on 29 July 1957. The Headquarters of the Agency are situated in Vienna. Its principal objective is "to accelerate and enlarge the contribution of atomic energy to peace, health and prosperity throughout the world".

IAEA NUCLEAR ENERGY SERIES No. NW-T-1.31

EXPERIENCE IN THE MANAGEMENT OF RADIOACTIVE WASTE AFTER NUCLEAR ACCIDENTS: A BASIS FOR PREPLANNING

INTERNATIONAL ATOMIC ENERGY AGENCY
VIENNA, 2022

COPYRIGHT NOTICE

Marketing and Sales Unit, Publishing Section
International Atomic Energy Agency
Vienna International Centre
PO Box 100
1400 Vienna, Austria
fax: +43 1 26007 22529
tel.: +43 1 2600 22417
email: sales.publications@iaea.org
www.iaea.org/publications

Printed by the IAEA in Austria
November 2022
STI/PUB/2022

IAEA Library Cataloguing in Publication Data

Names: International Atomic Energy Agency.
Title: Experience in the management of radioactive waste after nuclear accidents : a basis for preplanning / International Atomic Energy Agency.
Description: Vienna : International Atomic Energy Agency, 2022. | Series: IAEA nuclear energy series, ISSN 1995–7807 ; no. NW-T-1.31 | Includes bibliographical references.
Identifiers: IAEAL 22-01520 | ISBN 978–92–0–131122–1 (paperback : alk. paper) | ISBN 978–92–0–131222–8 (pdf) | ISBN 978–92–0–131322–5 (epub)
Subjects: LCSH: Nuclear accidents. | Radioactive wastes — Management. | Radioactive waste disposal. | Radioactive wastes — Decontamination.
Classification: UDC 621.039.7 | STI/PUB/2022

FOREWORD

The IAEA's statutory role is to "seek to accelerate and enlarge the contribution of atomic energy to peace, health and prosperity throughout the world". Among other functions, the IAEA is authorized to "foster the exchange of scientific and technical information on peaceful uses of atomic energy". One way this is achieved is through a range of technical publications including the IAEA Nuclear Energy Series.

The IAEA Nuclear Energy Series comprises publications designed to further the use of nuclear technologies in support of sustainable development, to advance nuclear science and technology, catalyse innovation and build capacity to support the existing and expanded use of nuclear power and nuclear science applications. The publications include information covering all policy, technological and management aspects of the definition and implementation of activities involving the peaceful use of nuclear technology. While the guidance provided in IAEA Nuclear Energy Series publications does not constitute Member States' consensus, it has undergone internal peer review and been made available to Member States for comment prior to publication.

The IAEA safety standards establish fundamental principles, requirements and recommendations to ensure nuclear safety and serve as a global reference for protecting people and the environment from harmful effects of ionizing radiation.

When IAEA Nuclear Energy Series publications address safety, it is ensured that the IAEA safety standards are referred to as the current boundary conditions for the application of nuclear technology.

Major accidents at a NPP or fuel cycle facility are rare but can produce large quantities of radioactive waste with widely varying characteristics that can be difficult to manage. This is illustrated by the challenges faced at the Three Mile Island accident in 1979 and the continuing challenges following accidents at the Windscale Pile No. 1 reactor in 1957, the Chornobyl NPP in 1986 and the Fukushima Daiichi NPP in 2011. Large volumes of radioactive waste can also be generated by accidents at military installations or the mishandling of high activity sealed radiation sources. The need to manage equivalent large volumes of waste from the cleanup of some legacy nuclear sites can also provide valuable experience for handling accident wastes.

Implementing safe, cost effective management of waste from a major nuclear accident has proven to be a complex, resource intensive undertaking. Substantial challenges can also arise in the case of smaller accidents. Decisions need to be made on the designs and technologies to be employed to treat and dispose of a diverse mix of waste constituents, and on the selection of sites to be used for predisposal waste management and the disposal facilities themselves.

In the case of a major accident, radioactive waste volumes can quickly overwhelm existing national management and disposal infrastructure. Appropriate disposal facilities might not be available to match the amounts or characteristics of the wastes. Under such circumstances, inappropriate response actions taken in the early aftermath of an accident can limit the range of future management and disposal options, substantially increase costs and result in significant worker exposures and increased risk of public exposure. Such situations may be avoided if precautionary accident response preplanning has been taken, even where the likelihood of serious accidents is considered extremely low.

Expanded knowledge from addressing wastes resulting from the Chernobyl and Fukushima Daiichi accidents, as well as experience with legacy nuclear fuel cycle and nuclear weapons facilities, non-nuclear accidents and radiological incident waste estimation tools offer valuable lessons for proactive preplanning and strategy development. Substantial experience has also been gained in many Member States in managing significant legacy radioactive waste in a suitably protective, cost effective manner using different approaches. This has improved operational approaches and provided experience in applying exemption and clearance principles.

Waste management challenges depend on the scope and severity of an accident, as well as on the stage of the response. The early emergency response stages may pose challenges when decisions need to be made quickly, there is limited personnel and facility availability, and the priority of the moment is controlling the emergency itself. Challenges can result from the collection and storage of small amounts of wastes in containers without special treatment; the need to create full scale conditioning systems to capture and stabilize radionuclides and damaged fuel; the requirement to manage a wide range of wastes; working in harsh physical and radiological conditions; and so on. Choosing and/or creating a waste management system is dependent on many factors, such as the amount of waste and its geographical distribution, levels of contamination, physical and chemical properties, techniques and resources available, and storage and disposal requirements and capabilities. Waste management can be a constraint to remediation if appropriate waste management facilities, logistics and staff support are not available in a timely manner. Integration of the remediation programme and the waste management programme is important.

Through robust preparedness planning, Member States can minimize the amount of waste requiring disposal, provide for separation of wastes by type and radioactivity level, and process or otherwise prepare stored waste for disposal and then dispose of it, all in a safe, efficient and cost effective manner. Achievement of these interrelated objectives is always conducted in a manner that is protective of workers, the public and the environment, in accordance with accepted standards. The experiences from past major accidents that are summarized in this publication are intended to support such preparedness planning.

The IAEA officers responsible for this publication were G.H. Nieder-Westermann and F.N. Dragolici of the Division of Nuclear Fuel Cycle and Waste Technology.

CONTENTS

1. INTRODUCTION

1.1. BACKGROUND

Large quantities of radioactive waste and/or wastes with widely varying characteristics are typically generated as a consequence of a major accident at a nuclear power plant (NPP) or fuel cycle facility. This is illustrated by the challenges faced with on-site waste management at the Three Mile Island (TMI)[1] Unit 2 accident in 1979 in Pennsylvania, United States of America (USA) [1, 2] and the continuing challenges with both on-site and off-site wastes following accidents at the Chornobyl NPP in 1986 [3–6] in Ukraine and the Fukushima Daiichi NPP in 2011 [7–12] in Japan. Large volumes of radioactive waste can also be generated by accidents at military installations or by unintentional/intentional acts involving dispersion of radioactive material in urban areas or other settings.

An accident at an NPP can result in a wide range of consequences, from essentially no direct environmental impact to significant releases of nuclear materials into the environment. This publication considers the lessons learned from waste management activities from the most severe nuclear industry accidents, focusing on experience from the Windscale Pile 1 reactor in Cumbria (formerly Cumberland), United Kingdom (UK), and the NPPs at TMI, Chornobyl and Fukushima Daiichi. Although these accidents had significantly different consequences, they are illustrative of the range of past accident impacts and can provide a basis for preplanning of waste management needs for future nuclear or radiological emergencies.

Significant releases of radionuclides into the environment occurred in the initiation and initial response phases of the accidents at the Windscale, Chornobyl and Fukushima Daiichi NPPs. At the Chornobyl NPP, fission gases, volatile radionuclides and fuel particles were released during the accident. The Windscale Piles and Fukushima Daiichi accidents resulted in the release of fission gases and volatile radioactive species. In contrast, the accident at TMI NPP resulted in little radionuclide release, essentially all noble fission product gases that were promptly dispersed.

In the case of the Fukushima Daiichi NPP, releases of mostly short lived radionuclides were spread to off-site territory, requiring extensive evacuation of the inhabitants and an active off-site cleaning campaign to reduce contamination and restore the human habitat to the greatest possible extent. The evacuated areas are being progressively repopulated at the time of writing this publication.

Efforts implemented after the Chernobyl accident included reduction of contamination levels and restoration of the human habitat to the greatest possible extent. However, in the Chornobyl NPP case, the off-site releases contained long lived actinides as well as short lived radionuclides, which resulted not only in evacuation of inhabitants, but also in establishment of the exclusion zone around the accident site. Due to the level and nature of the contamination, permanent habitation within the exclusion zone is prohibited for the foreseeable future.

In the case of the TMI accident, the effects were confined to the nuclear facility, without significant releases to the environment. In the Windscale Piles case, releases occurred, but mostly of very short lived radionuclides, with no need for evacuation of inhabitants or an off-site cleanup campaign.

Initiation of waste management steps will likely occur after essential emergency measures to stabilize the events are well under way or largely completed. However, emergency measures can also include partial cleanup of the site or affected facility to decrease radiation and contamination levels in order to gain or improve access to the facility, so some waste management activities will begin concurrently with emergency measures. These activities will likely involve segregation of the contaminated material by measurement of the gamma dose rate (GDR) and/or by material type (soil, vegetation, rubble, etc.) and collecting these raw wastes in temporary storage areas for easier implementation of predisposal steps. Priorities will be different for each accident case and will also differ between accidents with off-site releases and those where damage is localized/confined to within the nuclear facility or site. For example,

---

[1] A list of abbreviations is included at the end of this publication.

a priority emergency action in the case of the Chernobyl accident was primarily to decrease activity at the site so that workers at the multiunit NPP could continue to operate the three remaining units not seriously damaged during the accident. Off-site efforts were focused on decontamination to decrease activity in the environment, measures to control contamination spread and the establishment of controls for evacuated areas after the fallout. On the other hand, at Windscale Piles and TMI NPPs, a decision was made very early on to stabilize the situation by ensuring permanent shutdown of the production reactor and NPP, respectively. At Fukushima Daiichi NPP, the leading objective was to ensure stabilization of the three affected reactors and to shut down the other three remaining reactors permanently, even though Units 5 and 6 were not affected by the accident. In parallel, the huge efforts organized off-site to deal with fallout from the damaged reactors were complicated by a need to deal with the severe consequences of the natural disaster caused by the earthquake and subsequent tsunami. These different objectives resulted in far different approaches to handling the emergency situations, with subsequent impacts on implementation of waste management activities that are illustrated in this document.

The waste quantities produced following an accident can easily exceed the annual radioactive waste volumes generated within a Member State, overwhelming existing licensed radioactive waste management and disposal facilities. Such accidents are also likely to result in diverse, potentially problematic waste streams, not produced by routine operations. The physical, chemical and radiological characteristics of some wastes might not be compatible with existing treatment or disposal facilities. As a result, existing national infrastructure and procedures for radioactive waste management (waste characterization methods; facility operating procedures, equipment or design capacity; waste acceptance criteria for disposal; waste disposal facilities; transport and disposal permitting procedures; etc.) might be unable to cope with either the nature or the volume of the wastes generated.

The sudden nature of nuclear accidents can create crisis conditions due to the potential for immediate and significant risks to public health and safety, and widespread damage to property and economic activity. The crisis conditions can be magnified if the accident is caused by a widespread natural disaster — as was the case with Fukushima Daiichi — where the off-site infrastructure is severely damaged or otherwise challenged. Wastes from the initial phases of an accident are generated over a short period of time (days or weeks). Waste management is not the primary consideration during this initial emergency phase, which is rightly focused on preservation of human life and the environment. However, the actions and decisions taken in the early phases of the response may complicate future waste management steps. Following the initial emergency phase of the event, waste management activities could continue for a long period subsequently (years to decades). Careful preparedness planning will promote the preservation of beneficial management alternatives that might otherwise be constrained or foreclosed.

Waste streams generated by a nuclear or radiological accident can be different from waste generated by normal operations of nuclear facilities, comprising, for example:

— Large volumes of predominantly low contamination materials, up to millions of cubic metres, that might need to be treated as radioactive waste;
— Smaller quantities of uncontrolled high activity waste;
— Dispersed radionuclides not commonly found in wastes from normal operations.

The type and extent of contamination will be affected by the inventory and quantity of radionuclides in the facility at the time of the event and the inventory and quantity of radionuclides released by the damaged facility during the event, as well as the environmental conditions prevailing both during and subsequent to the event.

Waste characteristics are also dependent on the facility design, the event scenario, meteorological conditions, selective deposition of radionuclides, the decay of short lived radionuclides and the manner in and degree to which cleanup is carried out.

The nature of a waste management response to an accident involving the release of radioactive materials into the environment has to reflect both the scope of the event (e.g. the cause, type of facility, size of the affected area, etc.) and its severity (e.g. the mass, activity, half-life and rate of the release of radionuclides, their dispersion in the environment and proximity to population centres, and vulnerability of ecosystems). The severity of an accident would also be judged by any associated non-radiological impacts. The International Nuclear and Radiological Event Scale (INES) is a useful tool for communicating the safety significance of a nuclear or radiological accident or incident involving the release of radioactive materials into the environment. It is a logarithmic scale running from level 1 (lowest impact) to level 7 (highest impact).

Self-evidently, the waste management approach to an accident ought to be proportionate to its severity and scope. For relatively low impact events involving a limited radiological hazard in contained situations, radioactive materials might easily be collected and stored in simple containers while awaiting a disposal route. At the other end of the spectrum, the capture, stabilization and containment of high activity radionuclides or damaged nuclear fuel would likely involve remote or automated methods to limit exposure and collect a release, and sophisticated conditioning methods to allow for safe and secure storage until further decisions can be made. For any waste management situation, the details of the approach adopted will reflect a wide range of radiological and logistical factors, not least the volume of waste and its activity concentration, the physical and chemical characteristics of the waste, and the resources available to implement solutions.

In addition to the challenges of handling the accident impacts and the wastes being generated, there is a parallel challenge of maintaining effective communications with those affected. Most members of the public do not know what to expect if radioactive material is suddenly released into the environment, but they can be assumed to fear the worst. This situation can generate strong sociopolitical pressure to act quickly in response to both real and perceived dangers. Depending on circumstances, public and governmental interest may extend to neighbouring countries. Such conditions make tremendous human and financial resource demands on the organizations responsible for responding to the accident and accurately communicating what is being done. In many Member States, these responsibilities are divided among multiple organizations at the national, regional (state/provincial) and local level. This in itself presents daunting coordination, logistics and information sharing challenges.

It is recognized that it may not always be possible to observe ideal waste management practices, especially during the early stages of a nuclear accident, where bringing the situation safely under control is the primary objective. However, the waste management requirements and principles and their implications need to be considered at every stage of the accident, even if they are not completely achievable at the time, and full compliance needs to be sought within a reasonable time frame after the end of the initial accident phase. The basic safety principles for radioactive waste and post-accident management are included in other IAEA publications, as identified in Refs [13–20].

1.2. OBJECTIVE

Based on past experience of major nuclear accidents, several aspects of potential accident scenarios can be anticipated and used as a basis for advance planning and preparation that could significantly improve recovery in the event of a future accident [21].

The objective of this publication is to use this experience to provide systematic and comprehensive information on the technological aspects of managing the potentially large volumes and/or complex wastes generated over a short period of time during an accident. This information can be used to inform precautionary preplanning exercises that address possible accident scenarios for the nuclear facilities in a Member State. In this context, it is assumed that such wastes would be beyond the capacity and/or capability of the existing waste management infrastructure in a Member State to deal with.

Guidance provided here, describing good practices, represents expert opinion but does not constitute recommendations made on the basis of a consensus of Member States.

1.3. SCOPE

The primary focus of the publication is to provide guidance for technological planning and implementation of the management of wastes from nuclear accidents, from generation through to and including their eventual disposal, based on relevant experience from previous events.

Figure 1 provides a generalized chronology subsequent to an accident with respect to waste management. Representation of the activities as discrete phases of work is necessarily a simplification of actual situations. In practice, there will be significant overlaps between each of the activities. The urgency of the initial phase to control the spread of the consequences develops into an intermediate phase to make site conditions safer and then to a longer term phase for site remediation, decommissioning and restoration. The timeline illustrated is not to scale: for example, the initial emergency response phase may last from hours to weeks, while subsequent cleanup, decommissioning and remediation activities could last years to decades.

1.4. STRUCTURE

Section 2 describes the sources of waste that might result from an accident at a nuclear facility and their possible characteristics. Section 3 addresses the systems engineering approach to waste management, while Section 4 addresses pre- and post-accident planning. Section 5 discusses predisposal steps and options. Section 6 describes the waste management implementation process. Section 7 describes the various techniques that can be applied to the characterization of accident related wastes. Sections 8–11 discuss the collection/handling, transport, processing and storage of accident related wastes, respectively. Section 12 describes the options and procedures for disposing of the wastes. Section 13 provides summary conclusions. Each section begins with a short list of some of the key lessons that have been learned in the topic area from managing past nuclear accidents.

The six appendices provide brief summaries of the four major accidents and associated wastes that are used as key examples throughout this publication, plus information on other accidents and experience with managing large volume wastes from legacy nuclear site cleanup operations. Appendix I describes the 1957 accident at the Windscale Pile 1 reactor, Appendix II the accident at the TMI NPP, Appendix III the accident at Chornobyl NPP and Appendix IV the accident at Fukushima Daiichi NPP. Appendix V describes a selection of other accidents with lesser consequences and Appendix VI outlines some relevant work on managing the large volumes of waste (equivalent to the amounts that might result from an accident) that are generated during the cleanup of historic nuclear legacy sites.

This publication is one in a series of IAEA publications developed to support Member States' efforts towards improved preparedness in the event of a nuclear or radiological emergency:

- — IAEA-TECDOC-1826, Management of Large Volumes of Waste Arising in a Nuclear or Radiological Emergency [21];
- — IAEA Nuclear Energy Series No. NW-T-2.7, Experiences and Lessons Learned Worldwide in the Cleanup and Decommissioning of Nuclear Facilities in the Aftermath of Accidents [22];
- — IAEA Nuclear Energy Series No. NW-T-2.10, Decommissioning after a Nuclear Accident: Approaches, Techniques, Practices and Implementation Considerations [23].

The first publication, Management of Large Volumes of Waste Arising in a Nuclear or Radiological Emergency [21], provides wide perspectives considering safety, technology and societal aspects. A companion publication describes further technical experiences gained and lessons learned in the cleanup and decommissioning of nuclear facilities after an accident [22] and is supported by a further publication describing considerations related to techniques, practices and implementation methodologies [23]. Aspects relevant to the management of accident generated radioactive waste through to disposal are covered in the current publication.

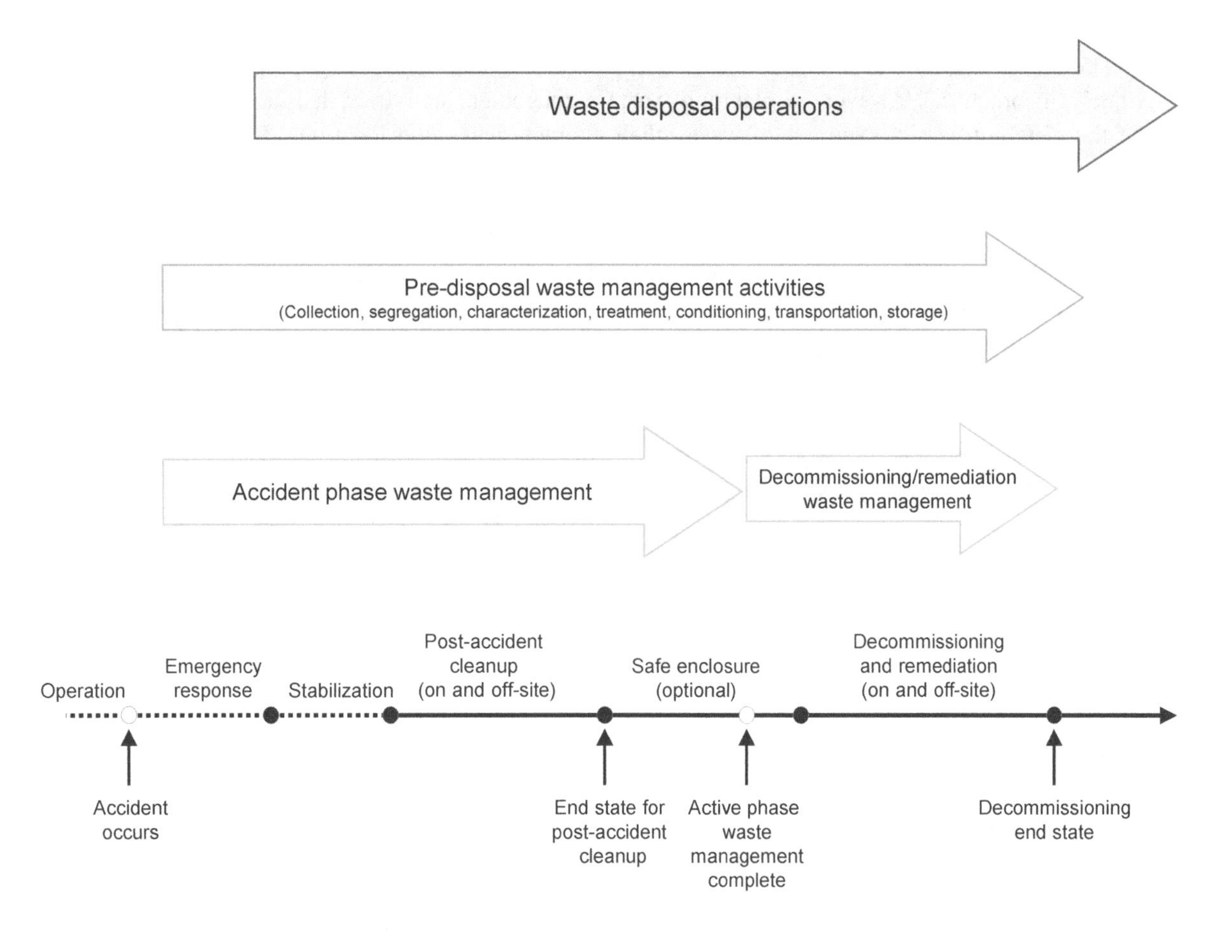

FIG. 1. Chronology relating to accident waste management. Adapted from Ref. [22].

The current publication addresses several sources of accident related wastes. For the purposes of this publication, these have been defined as:

(a) Primary waste: produced as a direct result of the accident (e.g. contaminated equipment, fuel debris, damaged or contaminated equipment or structures in the accident zone, etc.). Generally, these wastes will be solids located on-site and could have high activity.
(b) Recovery waste: produced as a result of stabilizing the conditions (e.g. water to cool a reactor) or making an area safe to enter (e.g. initial removal of highly contaminated debris). These wastes could be solid or liquid and will mostly be on-site.
(c) Restoration/remediation waste: generally solid wastes produced as a result of restoring the contaminated areas (e.g. soil, foliage, building rubble, etc.), including both on-site and off-site.
(d) Secondary waste: solid or liquid wastes associated with cleanup and other activities (e.g. workers' protective clothing, decontamination liquids, etc.) or treatment of primary wastes (e.g. off-gas filters from incineration, etc.).
(e) Incidental wastes: solid or liquid wastes produced as an indirect consequence of the accident, (e.g. cleanup of temporary storage areas after they have served their purpose), including both on-site and off-site.
(f) Decommissioning wastes: produced as a result of dismantling of the facility involved in the accident and any related facilities, including new treatment and storage facilities constructed or converted for use in the accident restoration/remediation. These wastes could be solid or liquid and will generally contain large equipment and/or building rubble. The wastes might have higher activity and/or volume than wastes from 'normal' decommissioning activities.

The above waste types might not be mutually exclusive and might not be segregated or managed separately in practice. However, in planning waste management activities, it is important to consider all of the different possible sources of waste, their characteristics and locations. In the context of this publication, 'waste' includes any contaminated material resulting from the accident, its cleanup and/or site restoration.

The publication does not address:

— Detailed considerations of irradiated fuel management (e.g. degraded fuel, corium, etc.);
— Nuclear material accounting and safeguards issues;
— Wastes from off-normal operation or minor incidents that can be handled with existing infrastructure and/or practices;
— Non-technological planning (e.g. government regulations etc.), which is covered in Ref. [21].

2. ACCIDENT WASTE ORIGINS AND CHARACTERISTICS

Key lessons learned:

— A preliminary and systematic estimation of total waste volume (classified first on the basis of activity level and subsequently classified based on type of waste) and waste generation rate during cleanup is required.
— It is almost certain that the wastes will be of larger volume and be more varied than 'normal' waste generated by an operating licensed nuclear facility.
— Planning has to consider that waste could be generated from both on-site (inside the boundary of the licensed site) and off-site (outside the site boundary) activities.
— The established cleanup criteria and anticipated end state will affect the quantities and types of waste generated.
— Management of radioactively contaminated hazardous (e.g. chemotoxic) waste may be challenging.

This publication addresses the management of radioactive waste generated as the result of a nuclear accident and subsequent cleanup/restoration activities. It does not deal with expected wastes generated by the normal operations of nuclear facilities. Nuclear accidents can generate extremely large volumes of heterogeneous, difficult to characterize waste. Examples of specific accidents and the generated waste streams used as the main basis for this publication are presented in the appendices. It is noted that this is not a complete list of historic accidents.

The IAEA Safety Glossary [24] does not define 'accident waste'. For the purposes of this publication, it is defined as radioactive waste arising from a nuclear accident that cannot be safely dealt with within the current framework of a national waste management system. All waste classes, as defined in IAEA Safety Standards Series No. GSG-1, Classification of Radioactive Waste [13], can be expected to arise, including exempt waste (EW), very short lived waste (VSLW), very low level waste (VLLW), low level waste (LLW), intermediate level waste (ILW) and high level waste (HLW). Wastes from different phases of an accident can also be defined, as described in Section 1. The general groupings of types and activities of wastes typically generated from a major nuclear accident are shown in Table 1.

The potential volumes of waste that can result from a nuclear accident are discussed in Section 2.1, waste characteristics are presented in Section 2.2, and the differences between wastes generated by the expected normal operation of nuclear facilities and those resulting from an accident are discussed in Section 2.3.

TABLE 1. TYPICAL TYPES OF WASTE RESULTING FROM A MAJOR NUCLEAR ACCIDENT

Waste activity level	Potentially large volume solid wastes	Large discrete items of solid waste	Liquid/wet solid wastes
LLW (can include EW, VSLW and VLLW, depending on national classification systems)	— Soil, both on and off-site — Debris, both on and off-site, e.g. trees — Concrete from a destroyed structure — Absorbers — Dust — Secondary wastes, e.g. used protective clothing	— Vehicles, etc. — Damaged equipment	— Water used for decontamination — Oils, solvents, etc. — Multiphase liquids
ILW	— Soil on-site — Debris on-site — Absorbers of radionuclides in coolant water, etc. — Incinerator ash — Concrete and other rubble	— Damaged equipment	— Coolant water of a reactor core — Decontamination liquids — Reprocessing facility raffinates — Sludge from purification systems — Spent ion exchange media
HLW	— Damaged fuel and fuel containing materials	— Reactor core	— Reprocessing facility raffinates

2.1. QUANTITIES OF WASTE

A preliminary and systematic estimation of total waste volume (classified first based on activity level and subsequently on waste type) and waste generation rate during cleanup is required for planning strategies to manage the potentially very large volume of radioactive waste generated by an accident. This information will form a key part of the basis for all further activities required in recovering from the accident. Considerations include:

— Cleanup criteria and action levels for decontamination (e.g. clearance level);
— Nature of the accident and type and scope of restoration (e.g. generation mechanism of waste, decontamination method and secondary waste, etc.);
— Volume and rate of waste generation estimated on a step by step basis relevant to the phase of the response (e.g. emergency phase, recovery phase, subsequent long term management period).

The types and quantities of radioactive wastes generated during the four major NPP accidents used as examples in this publication are described in detail in Appendices I–IV. These vary considerably from the lower end, such as at Windscale Piles and TMI NPPs, where tens of thousands of cubic metres of waste were generated, to Chornobyl and Fukushima, where millions of cubic metres of waste were generated. For comparison, the annual production of wastes from a typical 1000 MW(e) reactor is ~250–400 m^3 per year. Over a typical 60 year life this would amount to ~25 000 m^3, with another 10 000 m^3 resulting from decommissioning of the reactor.

In addition to these four well known major accidents, several other accidents have also generated large volumes of radioactive waste. These include (details of each are provided in Appendix V):

(a) The 1957 explosion of a storage tank with high activity liquid waste at the Mayak plutonium production plant near the town of Kyshtym in the Ural Mountains, Russian Federation;
(b) The 1966 crash of a military aircraft with nuclear weapons near Palomares, Spain;
(c) The 1983 accidental scrapping and melting of a teletherapy unit in Cuidad Juarez, Mexico;
(d) The 1987 mishandling of an abandoned radiotherapy unit by local residents in Goiânia, Brazil;
(e) The 1988 accidental smelting of a disused sealed radioactive source at the Acerinox foundry in Los Barrios, Spain.

There have been a number of accidents with limited or no direct off-site impacts that still had significant waste management consequences. For example, there have been several core damage events at sodium cooled reactors that required management of sodium bonded spent fuel and sodium contaminated systems and structures (Fermi-1, 1966; EBR-1, 1955), which presented technically complex waste management challenges for storage and disposal of the waste. Non-reactor accidents and events can present similar localized but serious waste management challenges. For example, the Hanford site (USA) plutonium uranium extraction plant rail transfer tunnels were retired and sealed off, with contaminated rail rolling stock interred. The tunnel and rolling stock were contaminated with nuclear materials from nuclear weapon production that had radiological, nuclear and toxicity safety implications. These required extensive and careful waste isolation and management actions.

2.2. WASTE CHARACTERISTICS

Detailed characterization, as described in Section 7, will be required at various steps of the management process in order to categorize the wastes for further management and identification of appropriate disposal solutions. The wastes resulting from an accident can usually be categorized as liquids, solids, wet solids or gases. The general properties of these waste streams are described below, and the key characteristics that need to be known for their management are summarized in Table 2.

(a) Liquids:
    - Large volumes; very low to high activity; complex chemical composition (e.g. salts, oils, suspended solids, biological fouling, etc.); can originate from many sources;
    - Examples include contaminated cooling water, natural sources (groundwater, rain, lakes/rivers/sea), decontamination water, etc.
(b) Solids:
    - Large volumes; very low to high activity; secondary wastes, materials brought into contaminated zone (tools, equipment, etc.); diverse materials;
    - Examples include damaged fuel, core debris, soil, rubble, metal, vegetation, components, protective equipment, contaminated dead livestock, contaminated human casualties, filters, incineration ashes, etc.
(c) Wet solids:
    - Low to high activity; primary and secondary wastes; diverse material characteristics;
    - Examples include ion exchange resins, slurries, sludges, etc.
(d) Gaseous:
    - Mixtures of fission product species (speciation dependent on facility type and operating conditions); varying radionuclide concentrations; potentially flammable mixtures;
    - Examples include volatile fission products from post-accident venting, flue gas from non-nuclear incinerators converted for radiological use (note: flue gas from incinerators can

contain volatile elements such as caesium, but bag filter systems can efficiently remove caesium to under detection limit, even if a high volume air sampler is used), etc.

For hazardous wastes (e.g. putrescible or chemotoxic materials), the non-radiological properties may be the dominant consideration. This can become relevant for processing/handling the wastes (e.g. where there is asbestos contamination), for storage of the waste (e.g. if the wastes display accelerated degradation) and eventual disposal (e.g. if the wastes contain lead).

2.3. DIFFERENCES BETWEEN ACCIDENT AND NORMAL WASTES

As already discussed, an accident can generate a large, unplanned volume of very heterogenous waste. To plan for management of these it is helpful to understand some of the important differences between accident wastes and wastes generated from normal operations/conditions. The major differences can be summarized as:

(a) The volumes of accident wastes are generally higher than those generated under normal operational conditions. For example, the volumes of liquids and solid wastes generated post-accident at

TABLE 2. KEY CHARACTERISTICS OF SOLID AND LIQUID WASTES RESULTING FROM A MAJOR NUCLEAR ACCIDENT

Characteristic	Solid and wet solid waste	Liquid waste
Physical characteristics	— Shape, volume and weight of wastes — Place of generation — Rheology (of sludges) — Liquid content	— Volume of wastes — Place of generation to help categorize the radiological properties
Chemical characteristics	— Chemical composition (metal, cement, polymer, complexes, etc.) — Chemical stability (e.g. self-reaction, reaction with oxygen, etc.) — Solubility of contaminants — Chemotoxicity	— Chemical composition (salinity, pH, organics, etc.) — Chemotoxicity
Radiological characteristics	— Activity levels of β–γ and α emitting radionuclides; — Content of fission products, activation products and actinides; — Equivalent dose level categories (e.g. <0.1, 0.1–10, 10–1000 and >1000 mSv/h): useful to determine appropriate preliminary temporary storage — Nuclear criticality issues (fuel residues)	— Activity levels of β–γ and α emitting radionuclides — Content of fission products, activation products and actinides — Nuclear criticality issues (fuel residues)
Thermal characteristics	— Flammability — Heat generation rate by radioactive decay and chemical reaction — Thermal conductivity	— Boiling point — Volatility — Flammability
Biological characteristics	— Biological reactions (e.g. decomposition, microbiological reactions etc.) — Putrescence of animal origin materials — Infectious agents	— Biological reactions (e.g. decomposition, microbiological reactions, etc.) — Infectious agents

Fukushima Daiichi NPP and the volume of solid waste generated at the Chornobyl NPP are many orders of magnitude greater that those generated during normal operations.

(b) The range of radionuclides and their concentrations in waste can be higher after a nuclear accident. For example, the quantities of short lived beta/gamma and fuel related alpha radionuclides released during the Chernobyl accident were vastly in excess of any gaseous releases that took place during normal operations.

(c) Short lived radionuclides may have substantially decayed by the time cleanup is under way. Therefore, they may be at lower concentrations than in normal wastes. This can be utilized in the planning process.

(d) Unlike with normal wastes, the chemical/biological content and hazards of accident wastes can be complex and variable.

(e) Waste characteristics, based on process knowledge, are known for normal wastes, but the nature of the wastes produced during an accident is dependent on the event scenario and the associated impacts, which are often not immediately known or fully understood.

(f) The physical characteristics of wastes resulting from accidents can be very diverse, leading to a wide range of waste types.

(g) Depending on the prevailing weather conditions, accident wastes (unlike normal wastes) can be scattered over a wide geographical range and may be distributed across national boundaries.

For liquid wastes, it is desirable to understand at least the following properties in order to plan for temporary storage and management, as well as the subsequent treatment steps, considering that the methods of treatment can differ according to the process by which wastes are collected:

(1) Activity concentration (e.g. radionuclide analysis);
(2) Place of generation (to help categorize the radiological properties);
(3) Waste generation volume and generation rate;
(4) Physical and chemical composition.

In addition, for a decontamination waste, it is necessary to understand the decontamination process and/or the agents employed.

In summary, accident wastes tend to be more voluminous, complex and variable than normal wastes and hence more difficult to characterize, segregate and assign to adequate management steps.

3. SYSTEMS ENGINEERING APPROACH TO WASTE MANAGEMENT PLANNING

Key lessons learned:

— Establish the roles, responsibilities and resources for developing and implementing the waste management strategy in advance and the authority for decision making.

— To achieve optimal solutions, the assessment and selection of waste management options has to evaluate all requirements against performance measures, taking account of their interactions and life cycle implications.

— A formal, systems engineering approach, implemented within a requirements management system, provides a robust framework for identifying and managing objectives, priorities and options, and could be developed in a national preplanning exercise.

— Develop a plan that facilitates communication with the public and other stakeholders.

— Some research and development as well as pilot scale testing may be required before designing and constructing full scale facilities.

A rational approach to post-accident waste management will ensure that the actions taken are effective and efficient. Priorities will change as the accident event unfolds and is stabilized. Waste management will not be the top priority in the early stages of the event. However, some of these early decisions could have a significant impact on the amount and type of waste that is generated. In addition, they could also limit or otherwise affect the options available for subsequent waste management steps. As such, it is considered good practice to establish in advance clear roles and responsibilities for developing and implementing a waste management strategy that can be deployed in the event of an accident occurring. In developing the strategy, it is prudent first to identify all technical and non-technical requirements that will be placed on the waste management system. These requirements will arise from many sources, both internally, from the organizations managing the accident and the facility concerned, and externally, from legal, regulatory, political and societal stakeholders, possibly at both national and international levels.

Experience has shown that the development and incorporation of transparent stakeholder engagement strategies can play an essential role in establishing public trust in the overall approach to waste management, thereby maintaining flexibility in subsequent response measures. To achieve success, it is necessary to work cooperatively and closely with regulators and stakeholders, to generate mutual trust and ensure that sound technical approaches will be identified, accepted and implemented, allowing cleanup efforts to be completed. It is important that the technical approaches are specifically selected to address and meet defined requirements.

The management of the TMI-2 cleanup provides an early example of the interplay of requirements, complicated by both technical and non-technical (e.g. regulatory, political, corporate, societal) challenges [2] Safety was placed above all other objectives and was used as the overriding requirement in planning and conducting activities. A decision not to restart the reactor was also made in the early stages and this helped focus waste treatment and decontamination activities. A strategy to pursue flexible and parallel cleanup technologies was also employed to avoid 'showstoppers' and minimize project restarts. The management team also worked with regulators and stakeholders to develop acceptable solutions. The regulations in the United States of America for low and intermediate level waste (LILW) disposal that were in effect at the time of the accident were neither complete nor explicit. Although more appropriate regulations were subsequently developed and issued, the TMI-2 management team at the time faced numerous technical and regulatory challenges, requiring extensive interaction with the regulator, as well as improvisation in the early approaches to waste management, particularly with respect to waste retrieval, storage and liquid processing systems. Nevertheless, radioactive waste management activities were conducted safely both before and after the promulgation of relevant regulations. Appropriate regulations, informed by the accident and response, were issued approximately three years after the accident.

In the autumn of 1979, a summary technical plan was developed to define performance objectives, set priorities and establish completion criteria [2]. In January 1980, approximately nine months after the accident, the Technical and Integration Office was established at the TMI. NPP This provided an on-site presence to direct cleanup activities [25]. In 1984, under the direction of the Technical and Integration Office, a programme strategy was developed that defined the cleanup plan in more detail. Figure 2 depicts this programme strategy.

Systematic identification of key objectives and priorities for waste management, coupled with an understanding of the most important drivers and constraints, and their implications for the available options, provides an important basis for event response and waste management decisions. The waste management strategy will be expected to meet numerous, sometimes competing, requirements. These will include requirements associated with the safety of workers on- and off-site, the safety of the public, nuclear safety regulations, security issues, the end states of the accident site and waste management facilities, local community needs, national and local political demands, international agreements, etc. If these requirements are not broadly understood before an accident and a strategy is not put in place swiftly to identify them, more specifically in the event of an accident, there are likely to be significant

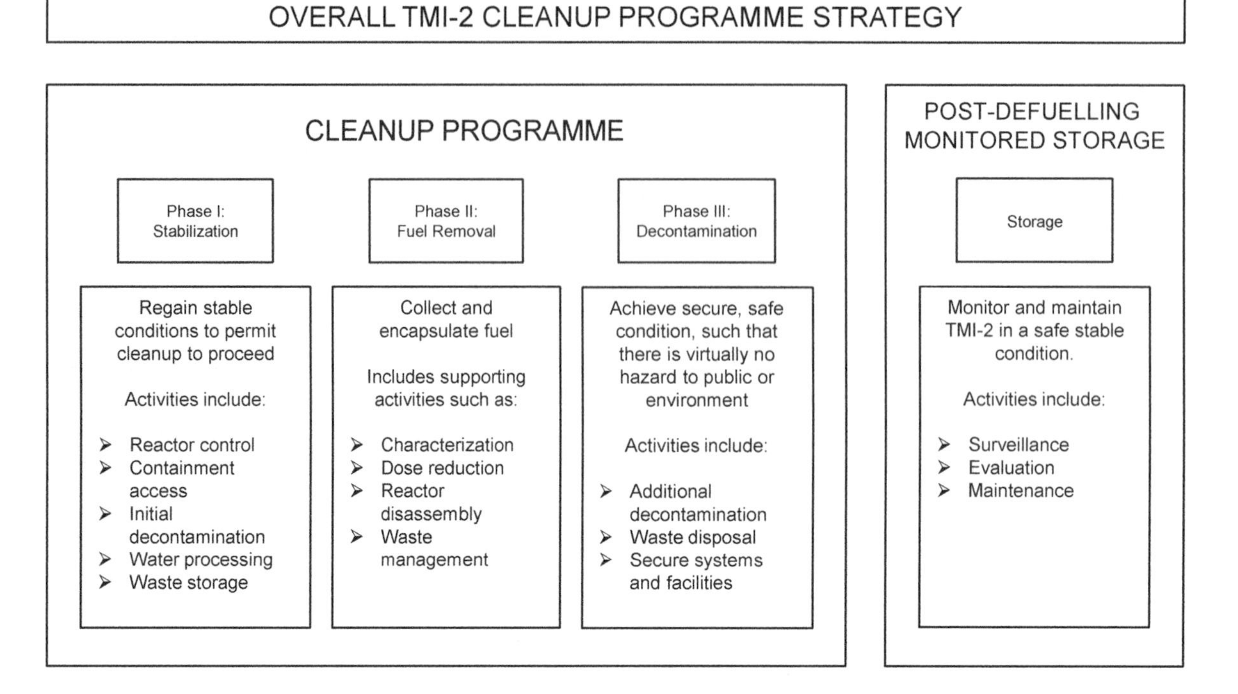

FIG. 2. TMI-2 cleanup strategy: phases and activities within phases. Adapted from Ref. [2].

downstream consequences. For example, technical solutions might be adopted that eventually fail to meet some key requirements (in particular to do with the properties of treated wastes and the location, design and end state of storage and disposal facilities), resulting in additional costs and radiological consequences in order to correct or adapt them.

A formal systems engineering approach, implemented within a requirements management system (RMS), provides a robust framework for identifying and managing objectives, priorities and options. With respect to accident preparedness, the incorporation of an RMS that defines the applicable requirements hierarchy and how their implementation can be demonstrated (Fig. 3) into a national preplanning exercise can be beneficial. In the event of an accident, the RMS can then be focused directly onto facility specific and accident specific circumstances. The use of the systems engineering approach in the context of waste disposal is described in a companion IAEA publication: Design Principles and Approaches for Radioactive Waste Repositories [26].

A formalized system engineering approach and a rigorous implementation of an RMS can be resource and cost intensive. Consistent requirements management involves several key elements, including requirements identification and determination of applicability; confirmation that implementation guidance and procedures are available for use; and provisions for confirmation that desired outcomes occur. Investment of time and resources to determine the effective requirements in the event of a range of accident scenarios that might occur in a Member State is advisable. It is important that the evaluation also considers relevant mechanisms for responding effectively to the requirements at all stages after an accident. It is noted that application of the systems engineering approach requires a serious commitment to its principles, a sustained implementation effort and realistic resource support.

3.1. REQUIREMENTS MANAGEMENT SYSTEM AND HIERARCHY

To achieve optimized solutions, the assessment and selection of waste management options will comprehensively evaluate, enumerate and manage the application of the salient requirements, including the use of performance measures. These can be structured following a hierarchical approach to managing

RMS Level		Description
1	High level external requirements	High level external requirements related to managing the wastes: international agreements, government policy, regulators, local community, etc.
2	Waste management system requirements	Qualitative and quantitative requirements that define how the total waste management system satisfies the high level external requirements : these would largely be defined internally by the organizations managing the facility and the accident response and would be used to define the response and waste management system.
3	Sub-system requirements	Specific requirements for each of the major structures, systems and components (SSCs) of the waste management system and the activities associated with them, where appropriate expressed as safety functions for each SSC.
4	Design requirements	Quantitative performance targets defined by the organizations managing the accident response and waste management system, for each SSC and activity, such that sub-system requirements and associated safety functions are met.
5	Design specifications	Performance targets for each component of each selected SSC design in the waste management system, plus detailed specifications for the design, construction, and manufacturing of each component or activity in order to meet the performance targets of the design requirements.

FIG. 3. Possible hierarchy of requirements in the RMS for an accident waste management system.

requirements. Figure 3 illustrates how an RMS for an accident waste management system might be constructed, based on a typical hierarchical structure that is organized into cascading requirement levels.

The highest level of an RMS principally needs to account for external drivers/requirements that set the boundary conditions for the overall approach to managing the wastes. These derive from legal, policy and regulatory sources, as well as strategic requirements from the organization responsible for managing the project. At this upper level, requirements will arise in the following principal areas:

(a) Regulatory compliance with respect to public and worker health and safety;
(b) Legal and regulatory compliance with respect to environmental impacts, both national and, possibly, trans-boundary;
(c) System feasibility, effectiveness and practicality requirements (including requirements for the use of available technologies, the use of available land areas and transport systems, compatibility between upstream and downstream processes, etc.);
(d) Cost and financing requirements (both overall and specifically identified for the different components of the planned approach, including requirements and constraints allowing initial costs to be traded off against costs for subsequent activities);
(e) Time schedule requirements (overall and detailed schedule objectives; lead and completion time requirements; availability of required resources and techniques);
(f) Public/stakeholder engagement and expectation requirements;

(g) Strategic requirements regarding the future of the accident facility and/or associated industrial complex.

As specific elements of the structures, systems and components (SSCs) to be managed by the waste management system are identified, system level and sub-system level requirements (levels 2 and 3) can be placed upon them, leading to design/process specifications at the final levels of the systems engineering RMS. The SSC would include existing components and structures within and external to the accident facility itself, plus new facilities/systems developed for managing the accident wastes. The lower levels of the RMS shown in Fig. 3 become progressively more specific and detailed, with the safety functions and performance targets of sub-system components and activities eventually leading to detailed design specifications.

3.2. IMPLEMENTATION OF A REQUIREMENTS MANAGEMENT SYSTEM

This section describes the application of an RMS strategy consistent with the hierarchical approach to requirements management described in Section 3.1. Specifically, the discussion focuses on the drivers of requirements that will need to be addressed in the RMS structure. These include:

— Accident site end state requirements and interim state milestones;
— Waste acceptance criteria for storage, transport and disposal facilities;
— Free release, clearance and recycling criteria;
— Discharge requirements;
— Licensing and regulatory requirements for all activities and facilities involved;
— Cost control and budget decision making;
— Waste minimization options and requirements;
— Records and data management requirements;
— Quality control requirements.

In addition, development of the RMS needs to take account of a range of constraints and opportunities that will affect the design and implementation of the waste management approach, including:

— Available organizational structures;
— Available infrastructure;
— Provisions for accommodating extraordinary costs that might be associated with accident waste management, such as the cost of facilities, material, support personnel and the capital cost of money;
— Post-accident demands on funding and funding sources, including indirect costs, such as converting assets to liquid capital to meet logistical and supply purchase expenses, etc.;
— Availability of personnel;
— Modes of deployment of technologies (fixed versus mobile/transportable);
— The possibility of interacting with other programmes or Member States on technical, logistical and financial needs;
— The ability to identify areas where research and development (R&D) would contribute;
— The need for continual feedback and recognition of lessons learned as activities proceed.

The lower level RMS requirement drivers and associated constraints and opportunities listed above are considered in more detail in the following subsections.

3.2.1. Accident site end state requirements and interim state milestones

Requirements defining the accident site end state are needed as early in the process as possible. The designation of the end state condition can dictate waste management technology decision making and deployment. For example, if an end state determination is made to allow only monitored future use in situ, deactivation and decommissioning techniques could be deployed, which could potentially simplify waste retrieval and waste treatment activities.

Alternatively, if the end state determination is to restore the site for general use, significantly more effort and more extensive treatment and technology will be required. Planning activities for remediation are therefore dependent on the desired end state condition. Policies and strategies for remediation of contaminated sites are addressed in other publications, such as Refs [16, 27, 28].

The determination of the end state requirement is not a straightforward exercise and may not be easy to achieve concurrently with competing priorities in the early aftermath of an accident. In addition, due to the long timescales that are typically required to reach a defined accident recovery end state condition, the application of an adaptive or phased approach to implementation of activities may be warranted. To this end, interim states with clearly defined milestones and interim phase end states that do not preclude the final end state determination can be defined. Definition of interim phase end states can also allow for increased flexibility in addressing changing stakeholder requirements, enhance public confidence and accommodate technological advancements, all of which may evolve over the timescales in question. Interim states can be useful in providing, among other considerations, time for a holistic approach to decision making and a more supportive and iterative stakeholder engagement process.

The New Safe Confinement (NSC) at Chornobyl NPP is an example of the use of such a phased approach. Installation of the NSC over the existing Shelter Object met the goal of providing a safe work area supporting future dismantling of the original Shelter Object and the eventual retrieval of the damaged core. The functional design life of the NSC is 100 years, to allow adequate time to develop, test and safely implement the required technologies, including the robotic and remote handling equipment needed for decommissioning the Shelter Object and Unit 4. The Shelter Object itself represents an interim state, designated to stabilize releases from Unit 4 and confine radiation to a controlled area.

A major consideration with respect to determining a final end state is the decision on whether to restart a damaged nuclear facility, as well as how a restart decision may affect undamaged adjacent units and facilities. The timing and impact of this determination will directly affect the focus of waste management, decommissioning, cleanup and remedial activities, versus repair and restart activities. The implementation of a dual track restart with cleanup will be an intense and complex management undertaking and will generate significant amounts of waste during and after execution. Experiences vary — for example, an early decision not to restart damaged units was made for TMI Unit 2 (TMI-2), Fukushima Daiichi Units 1–3, Chornobyl Unit 4 and Windscale Pile 1. However, decisions will be different regarding the use of other facilities at the nuclear sites where these damaged units were operated.

Because there were no releases outside damaged TMI-22, the decision was made to continue operation of TMI Unit 1 (TMI-1), located in its immediate neighbourhood. That set an interim state goal for the damaged TMI-2: to pursue a cleanup campaign that ensured safety conditions at TMI-2 could be met sufficiently to allow continued operation of TMI-1, which required several years. This also resulted in postponing decommissioning and eventual site remediation — that is, the end state condition — until the time when TMI-1 was shut down in September 2019. The demolition of the plant has been deferred, using the US Nuclear Regulatory Commission's (NRC's) SAFSTOR approach, permitting 60 years to achieve an end state, thereby also deferring the decision on the end state for the nuclear site to a later date. The approach will obviously affect the scale of waste management required.

The Chornobyl NPP was designed as a multiunit station where certain systems were shared between Units 1 and 2 and between Units 3 and 4. Units 1 and 2 were not adversely affected by the accident and isolation of Unit 3 from destroyed Unit 4 was considered possible. An interim state condition was defined: to confine radioactivity to the damaged unit by stabilizing the releases from Unit 4. This resulted in construction of the Shelter Object and initial cleanup goals within the exclusion zone , which allowed restoration of operations for several years at Units 1, 2 and 3. Ultimately a further decision was made to shut down all remaining reactors and, on 15 December 2000, Unit 3 became the last reactor to be shut down. In addition, designation of the exclusion zone allowed the containment of waste and the establishment of safe conditions for the affected population, permitting the cleanup response to move forward. The establishment of the exclusion zone and the evacuation of the population can be seen as an early interim state condition, supportive of ongoing and future decommissioning efforts.

When activity levels drop, the area can be redesignated. The Chornobyl exclusion zone could also be understood as a temporary location for long lived or HLW that will eventually be disposed of in a geological repository. The final decision on the end state of the plant site and the exclusion zone is still under discussion. There is an expectation, however, that the targeted end state will include future industrial use of at least part of the exclusion zone. An initial step towards this end has already been taken and a solar energy power station has been constructed within the exclusion zone.

At Fukushima Daiichi, a decision was reached not to restart undamaged Units 4, 5 and 6 and to focus on mitigation of the consequences of the accident for the whole nuclear site. This was consistent with the intention that the site will not be used for any nuclear activity in the future. This decision again affected waste generation and made it possible to locate temporary waste management facilities at the site itself. In the longer run the decision will of course affect the time frame for the decommissioning of all six units, as well as the total waste inventory that will need to be moved from the site to long term stores or repositories located off-site. Similarly, a fast paced off-site cleanup campaign, with a defined end state to restore most of the affected territory to the Ministry of Environment (MOE) defined future use conditions, will also affect the amount of waste to be generated, processed, stored, transported and disposed of.

Following the initial cleanup of the Windscale Pile 1 accident, the facility was placed under a regime of surveillance and maintenance. The reactor and support facilities are expected to remain in their current passively safe condition for a significant period of time, with full decommissioning currently planned in the 2040s. The current approach has been approved by the regulatory authorities and is subject to routine review.

As noted at the beginning of Section 3.2, a systems engineering approach is a proven optimization methodology through which requirements, in the form of objectives and boundary conditions such as those discussed above, are analysed via an RMS to identify a coherent strategy, priorities, implementation programmes and organization, in order to configure measures or projects. Figure 4 illustrates necessary inputs on how to determine a course of action for reaching decisions on end state and interim state conditions through the system analysis approach. As can be seen in Fig. 4, analysis starts with determination of the complex objectives that can be reached and understanding of the boundary conditions and available resources, as well of other limitations that are important in determining the course of actions.

3.2.2. Waste acceptance criteria for storage and disposal facilities

Storage and disposal facility waste acceptance criteria (WAC) need to be considered early and throughout the development and implementation of a waste management system and whenever transfer of the physical waste material or the responsibilities for waste material occurs, even if that transfer is within an entity's single organizational framework. WAC are an important form of requirement to be included in the RMS and are defined in terms of waste form characteristics. Waste form characterization is thus important to identify the radionuclides and radionuclide concentrations of the most important species and physical forms of the various waste streams that will require disposal, identify available solutions and determine the need for new disposal facilities. An initial broad picture of the range of waste form characteristics will help initially to scope possible management solutions: characterization can then be refined to create specific waste acceptance criteria suitable for each storage/disposal facility that is available or can be developed.

The characterization data can indicate solutions that simplify or eliminate waste storage needs for some waste types or, alternatively, drive the need to establish more extensive waste treatment or storage capabilities, either on- or off-site of the accident location. In some cases, they will indicate a need for new

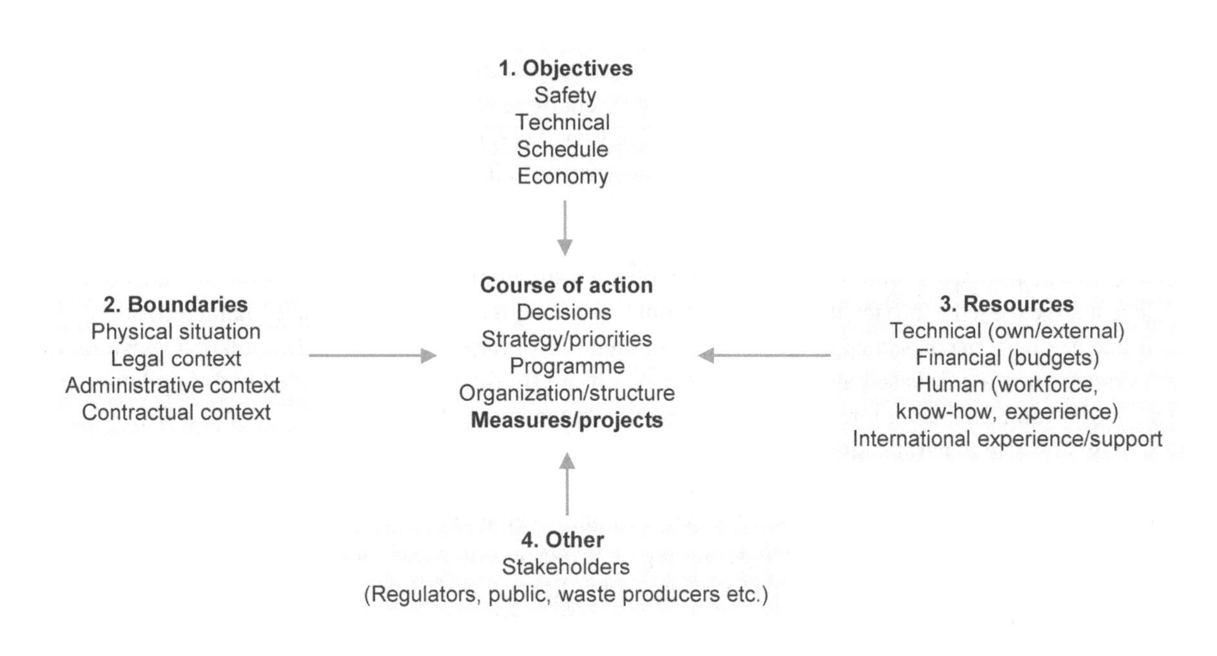

FIG. 4. Analysis of objectives and boundaries influencing decisions, configurations and performance of a programme or a project (system analysis method).

storage and disposal facilities and/or for the specification of new waste streams. In all cases, the WAC developed need to ensure that the feed streams are consistent with the capabilities of the initially available storage/disposal facilities, or those that will need to be implemented in the future. Waste retrieval and processing activities have to be conducted while being cognizant of the waste characteristics and waste acceptance requirements. Inappropriate actions could result in the generation of wastes that could challenge either storage or disposal or adversely affect the facilities.

In order to achieve alignment of the waste form characteristics, facility capability and WAC, the WAC have to be closely correlated with the requirements of the safety strategy and the findings of the safety assessment of the disposal facility, so that the choice and implementation of all predisposal (preceding) steps does not adversely affect storage or disposal. Both the Chernobyl and Fukushima Daiichi accidents resulted in waste forms, waste volumes and geographical dispersion of waste that were not previously anticipated. The actual conditions were well beyond any prior considerations of potential accidents and their consequences. The absence of relevant response preplanning caused the actual response activities to be far more reactive and, as discussed elsewhere in this publication, perhaps more rudimentary than would have occurred with a preplanned approach. Recognition that accident waste demands can, and likely will, exceed the wherewithal of existing storage and disposal capabilities is an inherent step in preplanning scenarios.

In identifying suitable disposal facilities, it is recognized that the current situation in many Member States is that the development of disposal facilities for wastes from NPPs is progressing slowly and the types of repository that would be needed for accident waste are not available. This is particularly the case for geological disposal facilities. Even where facilities do exist, the high volume, non-typical, problematic and complex waste streams from an accident will likely be difficult to accommodate, owing to space issues or non-conformance with the existing WAC of the facilities or with other requirements that form the (separate) RMS for the existing facility. Establishing new disposal facilities, both locally to the accident site and within an existing national framework of waste management, is likely to become a key challenge. Where a new storage or disposal capability is required at or near the accident site, relevant information from existing waste disposal facilities (e.g. with similar wastes, in similar geological environments, terrains or climates, nationally or in neighbouring Member States) may be available to inform both the design and development of the waste facilities and their WAC.

The extraordinary nature of the waste streams resulting from the accidents described in the appendices of this publication and their volumes and radiological characteristics make preliminary preplanning of feasible storage and disposal options and capabilities at a national and local scale prudent. The preplanning effort can consider a wide range of scenarios and will be valuable in providing the initial framework for waste management. If no preplanning has been done, or it has not been carried out at a sufficiently detailed level, then managing wastes in terms of either only partially applicable or yet to be specified WAC can present significant practical problems.

It is important to ensure that all parties involved in the development and approval of WAC fully understand the limits and conditions of such an unplanned situation and adjust their approaches accordingly to accept a risk based approach that will not compromise long term safety. Adequate characterization will facilitate a sound, risk based approach for the segregation, storage, treatment and ultimate disposition of the wastes. Such a risk based approach might prove more transparent and useful to both regulators and stakeholders and facilitate more cost effective waste management. As an initial approach, a first attempt to segregate waste materials after the accident can be made based on dose rates, physical characteristics and material types. Later, more detailed characterization will follow the requirements and conditions of processing, storage or disposal facilities that will be made available.

Implementation of various predisposal steps can be optimized to ensure the most cost effective disposal in appropriate facilities. Different steps of predisposal and WAC can be developed with such a goal in view. However, there may be situations when WAC for disposal in a given facility (e.g. a near surface facility) cannot be met due to long lived radionuclides or chemically toxic elements in the waste mix, or for other reasons. This can significantly complicate or increase the cost of predisposal processing steps. The fallback position, consisting of long term storage while identifying and locating an appropriate geological disposal facility for such types of waste, needs to be considered.

For new facilities, the development of waste acceptance criteria is an iterative process, closely related to design of the facilities (treatment, conditioning, storage and disposal) and their safety assessment and licensing, as depicted in Fig. 5, where the number of iterations is for illustrative purposes based on general design approaches for radioactive waste repositories [26].

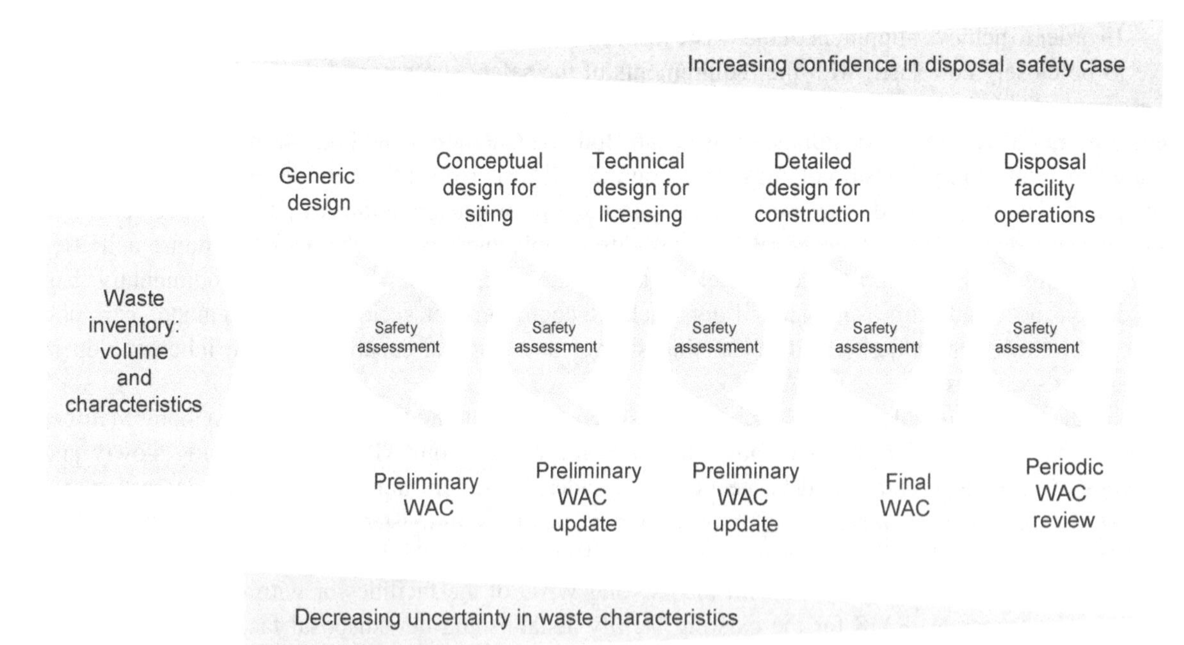

FIG. 5. Example of an iterative approach to the development of design, safety assessment and WAC.

3.2.2.1. *WAC at Three Mile Island NPP*

At the time of the TMI accident, waste classifications did not exist for all the solid waste that was generated from the TMI NPP cleanup. According to US regulations [29], LLW was classified as class A, class B and class C categories based on activity levels. Class A, B and C wastes could be disposed of in commercial facilities. It was realized that certain wastes exceeded US commercial burial limits (greater than class C — GTCC) so could not be disposed of using existing commercial burial sites. A programmatic environmental impact statement evaluation was conducted and an agreement was made between the US NRC and the US Department of Energy (DOE) for the DOE to accept these wastes for R&D purposes to evaluate treatment technologies, storage systems, etc. [30]. This part of the cleanup programme, known as the 'Abnormal Waste Shipment Program', became a vital element of the cleanup effort.

3.2.2.2. *WAC at Chornobyl NPP*

An exclusion zone was established after the Chernobyl accident due to the contamination of a large territory with long lived and alpha emitting radionuclides that currently cannot be remediated using cost effective measures. A decision was made to use the exclusion zone to locate all near surface disposal facilities for all waste streams from nuclear facilities and from the use of nuclear applications across the country, as well as long term storage facilities for waste that needs to await the development of a geological repository. The reference plan requires WAC for near surface facilities (NSFs) to ensure that after 200 years of operation and 300 years of post-closure monitoring, regulatory control can be removed from that disposal site. It is also assumed that waste from all new or existing storage facilities for HLW or LILW with long lived radionuclides will be removed to a suitable geological repository during the period of operation. With this approach, the WAC for NSFs have to be stringent, regardless of the fact that these near surface disposal facilities are located in the exclusion zone , which will need to be maintained in some shape or form for an undetermined time frame. If this continued long term exclusion were to be maintained, the timeline for the removal of regulatory control from NSFs might be extended. The additional time of regulatory control in consideration of radioactive half-lives could allow increased waste loading in conditioned waste packages, significantly decreasing the cost of disposal while maintaining safety considerations.

3.2.2.3. *WAC at Fukushima Daiichi NPP*

To evaluate potential on-site disposal concepts, existing safety assessments for several waste types can provide input into potential future disposal pathways (i.e. near surface, disposal at intermediate depth or deep geological disposal). However, at the early stage of waste management and in recognition of the uncertainties associated with the future development of disposal waste acceptance criteria, it is advisable that routine procedures are established for taking samples during waste collection and processing. These samples can provide the possibility for future detailed characterization, as needed to support disposal decisions. The disposition of bulk waste items generated by dismantling activities will be decided at a future date.

There are several criteria for contaminated wastes for off-site storage and disposal evaluation at Fukushima Daiichi NPP. A radiation level of 8000 Bq/kg was set when considering the radiation dose for workers at disposal sites (i.e. less than 1 mSv/year), while 100 000 Bq/kg was set when considering the radiation dose for residents near disposal sites (i.e. less than 10 μSv/year).

Contaminated wastes with >8000 Bq/kg, other than those generated by decontamination activities, are defined as 'designated wastes'. In Fukushima Prefecture, decontamination wastes and designated wastes with >100 000 Bq/kg will be sent to the interim storage facility (ISF) pending future disposal in a suitable engineered repository. Designated wastes with <100 000 Bq/kg will be disposed of in existing, controlled type disposal sites following procedures defined by Act on Special Measures concerning the Handling of Environment Pollution by Radioactive Materials.

Outside Fukushima Prefecture, it is planned that designated wastes will be disposed of in suitably engineered repositories controlled by the MOE. Disposal of decontamination wastes from outside of Fukushima Prefecture is still under discussion. All wastes with <8000 Bq/kg, except decontamination wastes, can be disposed of in existing controlled type disposal sites with other conventional wastes.

For wastes containing water-soluble radiocaesium, special attention is required to prevent rapid dissolution, which is achieved by covering them with non-contaminated soil. The Act prescribes methods for the disposal and control of contaminated wastes, including designated waste and wastes <8000 Bq/kg.

Further guidance on developing waste acceptance criteria can be found in other publications, such as Ref. [31].

3.2.3. Free release, clearance and recycling criteria

Large quantities of waste materials generated during a nuclear accident will have minimal levels of contamination. With adequate monitoring, they may be suitable for release or recycling for other purposes, rather than continued management as radioactive wastes. The policies, regulations and procedures on release of such materials vary among Member States and need to be included as high level requirements within the RMS. Release can be 'unconditional' (i.e. 'free release' for unrestricted use), or 'conditional' (e.g. for use within certain restrictions, such as for building of roadbeds, disposal in specified locations, reuse within specified industrial sectors, etc.). Beyond a clearance level (typically on the order of 100 Bq/kg), a radiation evaluation is required in every case, assuming many ways of radiation exposure (e.g. direct exposure, ingestion, inhalation, etc.).

At the Fukushima Daiichi NPP, contaminated materials were evaluated for use in the lower layer of road construction and material with <3000 Bq/kg was allowed to be used for this purpose [32]. According to this guideline, Miyagi Prefecture utilized bottom ash from the incineration of tsunami debris for landfill work. Corresponding to this regulation, the Japanese Geotechnical Society released a guideline for the recycling of contaminated incineration bottom ash [33]. For other materials, various dose limits have been regulated by the MOE [34].

At the Chornobyl NPP, free release of slightly contaminated accident material was not pursued due to the existence of the exclusion zone. Waste materials from on- and off-site decontamination were buried and marked. Initially, over 1000 earth trenches and mounds were established to contain contaminated material from the accident consequences liquidation campaign, carried out from 1986 to the end of 1988.

In addition, the official cadastre of Ukraine lists 53 decontamination waste storage facilities and special decontamination stations outside the exclusion zone that were used mostly for the decontamination of vehicles used inside the exclusion zone during the liquidation campaign. These facilities are maintained by Kiev 'Radon' and are situated in the Kiev Zhytomir and Chernigov regions.

Currently, based on investigative results, approximately 200 trenches within the exclusion zone have been reclassified as non-radioactive waste sites. A similar effort is ongoing for the decontamination waste storage facilities and special decontamination stations outside the exclusion zone.

The radioactive waste management programme at the Chornobyl NPP site provides for the operation of a facility for the release of waste from regulatory control. Work specific to this facility is contracted under a European Commission funded project.

In general, it is noted that strong public or political sentiment can be expected that will influence decisions on clearance levels and release and recycling of contaminated materials from a major accident. These sentiments, some of which might be formally identified as external requirements in the RMS, might oppose implementing clearance or at least restrict application to within the contaminated zones or affected areas. Implementation of clearance requires derivation of criteria for release (e.g. maximum radionuclide concentrations) as well as measurement protocols to confirm that the materials are below these levels. The measurements require sensitive equipment to measure the very low radiation levels adequately. Further guidance on measurement techniques for application in clearance is given in Section 7. Further guidance on free release, clearance and recycling can be found in other publications, such as Refs [35–37].

3.2.4. Discharge requirements

Disposition of some treated or untreated liquid or gaseous wastes streams may be accomplished by discharge to the environment, made in compliance with national regulations and within authorized discharge limits, which will form specific requirements at a high level in the RMS [38, 39]. An understanding of discharge limits is imperative to guide treatment technology selection and operations, and to determine go/no go decisions for candidate technologies. It may be necessary for the implementing agency to work with regulators with respect to agreeing on case specific pathway analyses, while always respecting the intention of the safety regulations and public sentiment. For example, very low concentrations of tritium that cannot be removed from aqueous streams might require flexibility in disposal decisions to facilitate cost effective management.

As a result of the Windscale Piles accident, the hazard associated with the release of fission products into the wider environment was recognized. While inhalation and external radiation risk appeared to be limited, measurements of the iodine content in locally produced milk identified a more immediate concern, requiring emergency measures. Consequently, the collection and distribution of milk over a 500 km^2 area around the site was restricted. Milk was disposed of via wastewater drains to avoid ingestion while ensuring adequate dilution in the environment.

The emergency measures adopted by local authorities (formerly Cumberland, now Cumbria) were examined for their effectiveness after the event. One fundamental weakness readily identified was that there was no published information at the time concerning a ‘safe’ level for ^{131}I in milk. As a result, a highly cautious approach was taken, banning the consumption of milk over a relatively large area. It was also apparent that the time between the accident and the initiation of the milk sampling programme was longer than would otherwise have been the case if effective procedures and resources had been planned for in the event of an accident. Following the incident, there was some criticism for not making the area pessimistically large at the beginning, and then shrinking it, since it was considered that this approach would have been psychologically more acceptable. The United Kingdom Atomic Energy Authority (UKAEA) stated that it would pay compensation for milk disposed of [40].

The nature of a waste management response to an accident involving the release of radioactive materials into the environment needs to reflect the scope of the event (e.g. the cause, type of facility, size of the affected area, etc.) as well as the severity of the event (e.g. the mass, activity, half-life and rate of the release of radionuclides, their dispersion in the environment and proximity to population centres and vulnerability of ecosystems). The severity of an accident would also be judged by any associated non-radiological impacts. The International Nuclear and Radiological Event Scale (INES) is a useful tool for communicating the safety significance of a nuclear or radiological accident or incident involving the release of radioactive materials into the environment. It is a logarithmic scale running from level 1 (lowest impact) to level 7 (highest impact).

Self-evidently, the waste management approach to an accident ought to be proportionate to its severity and scope. For relatively low impact events involving a limited radiological hazard in contained situations, radioactive materials might easily be collected and stored in simple containers while awaiting decay, long term storage or a disposal route. At the other end of the spectrum, the capture, stabilization and containment of high activity radionuclides or damaged nuclear fuel would likely involve remote or automated methods to limit exposure and collect a release, and sophisticated conditioning methods to allow for safe and secure storage until further decisions can be made. For any waste management situation, the details of the approach adopted would reflect a wide range of radiological and logistical factors, not least of which the volume of waste and its activity concentration, the physical and chemical characteristics of the waste and the resources available to implement solutions.

Due to the nature of the accident, the solid wastes generated from the Windscale Piles accident were localized to the site and, although there were larger than normal generation rates, these did not overwhelm the existing infrastructure. These wastes were disposed of at existing facilities or retained in a passively safe state within the reactor enclosure.

Approximately 7000 m^3 of water was pumped into the reactor during the accident. The water mainly went into the pile cooling pond, with some overflowing to the air ducts. The water from the pile cooling pond was discharged to sea (approximately 1.1 $m^3 \times 10^5$ m^3) and the water from the air ducts into the surface drainage system. By the time the water was discharged to the sea, over a period of a month, it was much less active as a result of radioactive decay [41].

As another example, the immediate releases from the TMI accident were largely limited to noble fission product gas (Kr), released to the coolant as a result of the core damage and then stripped from the coolant by the plant's systems in such volumes as to result in unplanned releases up the plant stack as relief valves actuated. Even considering the stable weather conditions that existed for most of the release durations, the calculated maximum doses to the public off-site were very small 'immersion' doses of the order of 10 μSv. During the initial response and recovery stages of the accident, no exposure to significant radioactive particulate releases or solid or liquid radioactive waste discharges occurred off-site [42].

At TMI NPP, disposing of water after treatment became a significant issue, owing to the remaining tritium concentrations. Tritium concentrations were at a low enough level that direct discharge would have provided sufficient dilution for them to be well below regulatory limits. However, due to political sensitivities and lack of public acceptance, an evaporation approach was finally implemented [2]. There are similar ongoing issues related to water discharge at the Fukushima Daiichi NPP, where large quantities of treated water are being stored because of residual tritium

3.2.5. Licensing and regulatory requirements for all activities and facilities involved

Existing regulations and licensing requirements have to be acknowledged throughout the waste management system and be incorporated into the RMS. Establishing a working relationship with regulators and stakeholders is important to ensure mutual understanding of requirements. Applicable international standards, regulations and guidelines — including Refs [13–20, 35–39, 43–49] — may facilitate licensing and provide valuable information to support regulatory discussions. The unique nature of the wastes after an accident could require special consideration by the regulatory bodies. For example, an important element of the TMI NPP cleanup effort was the Abnormal Waste Program [1]. Under this programme, the US DOE agreed to accept certain wastes that exceeded limits for commercial burial. This provided a solution for the disposal of several waste streams (e.g. zeolite water treatment media) that could not be provided for under existing regulations. This illustrates the cooperative relationship that was established between the commercial operator of the NPP and the federal government to provide a rational approach to solve a difficult waste management problem.

3.2.6. Cost control and budget decision making

Due to the large volumes and complexity of wastes resulting from a nuclear accident, waste management costs can be exceptionally high. Mechanisms for the acquisition of sufficient funding and for optimizing spending and controlling costs have to be considered comprehensively for the waste management system, and preplanning again plays a significant role. The systems engineering approach discussed in this section can facilitate option analysis and decision making processes to inform facility owner/operators, governmental appropriators and regulators. While the primary decision makers are likely the facility's executive team, decision making and/or the execution of the decisions can be significantly affected by external stakeholders. Identifying appropriate budgetary and spending requirements in the RMS will be a sensitive issue that can be explored beneficially in preplanning scenario exercises.

As part of the cost management and funding, the facility executive team can provide for effective and conservative cost estimating, aggressive cost and schedule management, and sufficient budget contingency/reserve. This planning and engagement approach could build confidence in the cost management and funding needs for involved parties. Though varying between Member States, the involved parties typically and necessarily include both policy and funding decision makers in governmental bodies, multiple regulatory organizations, insurance underwriters and peer owner/operator resource support.

For example, shortly after the accident at TMI-2, the US Government collaborated in the recovery by providing funding for various aspects of waste management, including the long term storage of core rubble, decontamination technology and reactor core technology data collection. The funding, based in significant part on the utility cost estimates, provided a cost offset, reducing the gross expenditures of both the utility and its underwriters [50].

As part of the planning and budgeting process, cost–benefit–funding impact analyses can be utilized to support technology planning and decisions. Waste management requires advance assessment and planning to ensure that the most cost effective option that satisfies requirements is selected. The assessment process typically involves defining requirements and objectives, identifying options and scenarios for implementation, defining evaluation criteria, evaluating options and selecting the preferred option. Multiattribute analysis tools are also essential tools in better understanding potential alternatives. The attributes to be evaluated include:

- The waste management goal;
- The types and quantities of waste and their locations;
- Versatility to handle different or new waste types;
- Risks related to worker safety as well as to long term safety of the waste packages that are important for disposal;
- Time necessary for implementation;
- Costs of the alternatives considered.

Inclusion of formal change control procedures within the systems engineering approach allows options to be identified and evaluated in order to adapt to, for example, the introduction of new, problematic waste streams, or other changes. This results in a decision that can be justified based on a sound decision making process. Further information and guidance on applying a multiattribute decision making process to the selection of waste management options can be found elsewhere, such as in Ref. [51].

3.2.7. Waste minimization options

As with all waste management systems, waste minimization is an important component of accident waste management. The quantity and type of waste generated during cleanup and remediation activities are governed by the radionuclides and their concentrations, the degree of contamination spread and the chosen site end state and disposal options [52]. The degree of contamination spread mainly depends on the nature of the accident and the prevailing environmental conditions at the time of the accident (wind, precipitation, surface water flows, etc.). These specific conditions are beyond the control of any operating or accident management organization. However, it is advisable to study one or more beyond design basis, more extreme scenarios of possible environmental conditions specific to a particular site in order to better consider and prepare for the potential consequences of such an accident (as the experiences from the Chornobyl and Fukushima Daiichi NPPs would suggest). Such additional studies and analysis of potential consequences may be valuable to consider for integration into the safety case for operating or new sites. The better understanding achieved for very low probability consequences does not automatically result in the establishment of new boundary conditions or a new set of engineered barriers, but rather provides for better preparation for the management of accident consequences, including the management of resulting waste.

In addition, large scale evacuations may also contribute to an uncontrolled spread of contamination (e.g. by vehicle movements, contaminated personal property, etc.). Controlling this spread of contamination is a responsibility of the accident response management organization, although it may not be their top priority (which is the immediate health and safety of the affected people).

Another aspect of waste minimization that is often overlooked is the eventual decommissioning of waste from storage and handling facilities that have been created to treat the accident wastes. Each facility will itself become waste at the end of its life. Therefore, it is necessary to plan the number and nature of the created facilities carefully and, to the extent possible, optimize these with respect to decommissioning waste. Further, it is important that, where practical, the number and types of facilities and the equipment located in the facilities be consistent with applicable safety and environmental protection requirements. Schedules and set deadlines are to be avoided as drivers, unless they are driven by worker or public protection needs.

It is advisable that in the design of any new facilities allowance be made for easy decontamination and minimization of secondary wastes. Prudent consideration and documentation of key radiological conditions are needed, with record preservation requirements extending through the operations phase in order to support decommissioning and decontamination. Such records can include routine surveys of operating areas; periodic surveys and sampling of hoppers, tanks, high integrity containers (HICs), etc.; records of spills, inadvertent releases and contamination outside normal radiologically controlled areas; and similar records for both normal and anomalous radiological contamination.

3.2.8. Records and data management

It will be necessary to establish a records management system early on to support the waste management effort. Waste related records may be required for many decades, so the records management system needs to consider the degradation of physical records or the obsolescence of digital systems [53].

Operational records of site and near site conditions prior to and during the accident phase can be valuable in waste characteristic assessments. Additionally, records kept during the cleanup phase are needed to understand changing site conditions and support waste forecasting. Characterization of waste streams is needed to determine the capability requirements for treatment, storage and eventually disposal. Management of the characterization data is needed to accurately support future treatment, storage and disposal requirements resulting from changing site conditions. Waste forecasts will need to account for existing wastes, as well as wastes that will be newly generated during future activities, and secondary wastes resulting from planned waste recovery and treatment/processing activities.

The need for data and records management from the inception of the accident was an important lesson learned from the TMI accident [2]. Initially, emergency response dictated work activities, with little emphasis on assessing and documenting plant conditions. Once the plant was stable, focus turned to cleanup, leading to the ultimate goal of defuelling in the fastest manner possible. It was learned that data acquisition and interpretation were vital to decision making. Therefore, a key lesson learned was the need to centralize data collection, management and analysis functions. Further, controlling how data were gathered was needed to ensure data consistency, quality and, to the maximum extent, comprehensiveness. Another important observation was the need to thoroughly understand the objectives pertaining to sample analyses, including understanding the analytical capabilities to be used (and their limitations) and ensuring clarity of data presentation (e.g. units).

Data management was also recognized as being important in the Fukushima Daiichi NPP case. Details are described in the guideline for the pilot transfer programme. New data management systems have been developed by the Japanese general construction companies involved in the cleanup.

Further guidance on waste related records and data management can be found in other publications, such as Refs [54, 55].

3.2.9. Quality management challenges

Nuclear accidents present significant quality management challenges. Preplanning to develop quality management systems in anticipation of potential problems can help to avoid adverse impacts that may arise due to the sudden, complex and rapidly evolving nature of most nuclear accidents. By identifying significant risk areas, systems and processes can be placed in readiness before an accident occurs. Risk areas relevant to waste predisposal and disposal include:

- — Development of new disposal facilities for all accident wastes could be significantly delayed or prevented if there is insufficient information on the wastes or on facility performance, hindering the completion of the necessary safety assessments;
- — Inability to provide acceptable data to meet WAC may prevent receipt at existing treatment or disposal facilities;
- — Increased worker exposure to radiological and other safety hazards might result if waste constituents are not known, or incompatible wastes are combined;
- — Poor control of processes that result in mixing of lower concentration wastes with highly contaminated wastes can lead to large quantities of waste needing to be disposed of according to the highest risk constituents, at much greater cost than may otherwise be the case.

The absence of an acceptable quality management system can also lead to other waste predisposal and disposal problems, including:

- — Insufficient confidence in the engineering, procurement, construction and operation of waste management facilities due to absent or weak quality assurance, which will have a direct impact on public support for the national accident waste disposal strategy;
- — The use of non-qualified workers if the quality of training is inadequate.

3.2.10. Available organizational structure

An organization is needed to establish, govern, coordinate and execute waste management activities. It is important that the organization has clearly defined responsibilities with accountability to the stakeholders and regulators, and well defined interface relationships with the accident response team. The organization's governance approach needs to include typical functions such as strategic decision making and direction for execution, project management and project controls, but additionally, the organization needs to include or possess the ability to influence health and safety, technology development, training and procurement functions. The organization and the response will need to remain flexible as the organization will evolve and will need to adapt to changing demands and conditions over the course of the response, which could take years to decades.

3.2.10.1. Windscale Piles

Throughout the Windscale Piles accident and during the recovery phase, the site, UKAEA and local community organizational structures (e.g. local constabulary, milk marketing board) in place at that time were maintained. Although initially a small team of eight people began to discharge the reactor, the response team grew to include many individuals from the site workforce, along with specialist resources drafted on a volunteer basis from other sites. The investigation of the accident itself led to the creation of a licensing board for all future civilian nuclear reactors. In 1959, the Nuclear Installations Act was passed, setting in motion the formation of the Inspectorate of Nuclear Installations within the Ministry of Power. These organizations were chartered to provide collectively the required governance and support functions for all civilian reactors.

Similarly, decommissioning activities, under the oversight of the inspectorate, were implemented on Piles 1 and 2, and structured project frameworks were developed to coordinate, communicate, align, manage and control all activities, ensuring that progress remains aligned to strategic objectives and the project delivers a credible plan. The Sellafield site is now owned by the Nuclear Decommissioning Authority and Sellafield Ltd is responsible for decommissioning of the Windscale Piles. The plans call for the removal of the fuel from the piles by 2030, with full decommissioning to be completed in the 2040s.

3.2.10.2. Three Mile Island NPP

The organizational structure at TMI NPP evolved from the plant operations organization in place immediately after the accident, through organizations more adapted to the cleanup and waste disposition. In some cases, this change worked well to fit the needs of the activities at hand. However, the changes were a strain on personnel as a result of recurring changes in their roles and the need to adapt to new organizations.

The first organization was structured for crisis management and served well in stabilizing the plant after the accident. Changes to departmental structures in the organization reflected the emergent needs of the cleanup. This adaptive structure evolved into a project oriented organization as cleanup activities were initiated and executed, requiring a shift in focus to management of the stabilized but heavily damaged facility, and to management of the accident, recovery and initial cleanup activity wastes. There were significant organizational challenges throughout the programme as multiple contractors were integrated and unplanned events required adaptive responses. The latter was especially true during the defuelling stage and required the formation of task groups that cross-cut organizational lines to solve emergent issues.

By the early 1980s the majority of fuel had been removed from Unit 2, packaged and transported to Idaho National Laboratory for storage. Approximately 1% of the fuel remains in inaccessible parts of the reactor vessel. Unit 2 was transferred to FirstEnergy Corporation and put into a safe storage and stable condition pending future decommissioning. The recovery organization evolved again into the much smaller organization needed to monitor the unit in a stable long term storage configuration. The Unit 2 licence was transferred to a decommissioning contractor in December 2020 [56]. Unit 1, owned by Exelon Corporation, remained operational until September 2019, when the reactor was permanently shut down. It is planned to place Unit 1 into a safe storage condition with decommissioning delayed, perhaps until 2075.

3.2.10.3. Chornobyl NPP

During the Chernobyl accident, immediate emergency measures were directed by a special team of state experts and led by specialized units of the former Soviet Union's military. This organizational approach dealt with the immediate accident consequences, such as firefighting, cooling down the reactor remains and evacuation of the population, followed by creation of the exclusion zone. Accident recovery work commenced immediately after the accident in April 1986 under the supervision of the State Commission of the USSR [57], which continued its activities until 1991, and the organizational structure evolved to address monitoring and management of both the local population and the extensive property affected by the contamination.

The 2600 m^2 exclusion zone was established shortly after the accident in May 1986 and placed under military control, with the population evacuated owing to the levels of radiation and the risk of contamination. The government refocused all activities of the Academy of Sciences of Ukraine and other state institutions and organizations onto providing technical and scientific support to the government's accident response. Within a week of the accident the government established the Operations Group (OG) to coordinate various response efforts.

The Academy of Sciences, the Ministry of Water Resources, the State Agricultural Department of Ukraine and other involved agencies, set up an analytical centre at the Institute of Cybernetics of the Ukrainian National Academy of Science tasked with assessing possible contamination along the course of the Dnipro River. In autumn 1986, the first forecast was presented to the OG and the Ukrainian government, and regular forecasting continued until 1998. A characteristic feature of the activities of all official commissions and, primarily, the government OG during this period was close cooperation with scientists.

As a consequence of the accident, an interdisciplinary commission was established in 1989 through the Academy of Sciences of Ukraine, tasked with augmenting draft laws on protection of the population. The basic principles of the laws were developed by researchers from Ukraine working jointly with colleagues from Belarus and Russian Federation. This work served as a basis for the Verkhovna Rada (parliament) of Ukraine to adopt more responsive laws and regulatory legal documents that considerably relieved the social and economic stresses amongst recovery workers and the affected population.

Today, the State Agency of Ukraine on Exclusion Zone Management (SAUEZM), as part of the Ministry of Ecology and Natural Resources, is entrusted with implementation of the state policy for radioactive waste management, including the long term management of radioactive waste storage and disposal sites. To perform its functions, SAUEZM works with the state owned specialized radioactive waste management enterprises at the Chornobyl NPP responsible for decommissioning Units 1–3 and the transformation of the Shelter Object into an environmentally safe system. In compliance with state policy, waste generators are responsible for management of waste prior to transfer to specialized radioactive waste management enterprises. The waste generators are precluded by law from disposing of radioactive waste.

3.2.10.4. Fukushima Daiichi NPP

At the time of the accident, the Fukushima Daiichi NPP was owned and operated by Tokyo Electric Power Company (TEPCO), under the regulatory oversight of the Nuclear and Industrial Safety Agency (NISA), a special organization attached to the Agency for Natural Resources and Energy under the Ministry of Economy, Trade and Industry. One of the roles of the Agency for Natural Resources and Energy was to promote nuclear power. The Nuclear Safety Commission, under the Cabinet of Japan, had review authority for work conducted by the regulatory agencies, such as NISA.

In the immediate aftermath of the earthquake, the Government established the Nuclear Emergency Response Headquarters, which issued the necessary evacuation orders. NISA also established its Emergency Response Centre, composed of relevant Ministries, including the Japan Self-Defense Force and other relevant organizations, such as the Japan Nuclear Energy Safety Organization (JNES, which provided technical support to NISA), the Japan Atomic Energy Agency (JAEA, R&D organization), etc. and started collecting information on the reactors in the affected area.

Following a full review of the safety regulation in the aftermath of the Fukushima Daiichi accident, the Nuclear Regulatory Authority (NRA) was established as an external agency of the Ministry of the Environment. The Nuclear Safety Commission and NISA, including the regulatory authority of the Minister of Economy, Trade and Industry, were transferred to the NRA. JNES was also absorbed by the NRA.

In September 2011, the Nuclear Damage Compensation Facilitation Corporation was established under the Nuclear Damage Compensation Facilitation Corporation Act to ensure that compensation payouts would be promptly and appropriately provided [58]. The goal was to ensure a stable supply of electricity by granting compensation funds required by nuclear facility operators potentially faced with large scale nuclear damages.

As all reactors at the Fukushima Daiichi site reached cold shutdown status in December 2011, the government adopted the Mid- and Long-Term Roadmap towards the Decommissioning of TEPCO's Fukushima Daiichi NPP [59] and the decommissioning phase began based on the roadmap. The roadmap has since been updated five times; the most recent update was adopted in December 2019. The International Research Institute for Nuclear Decommissioning (IRID) was established in August 2013 to conduct R&D on technologies that could be used for decommissioning the NPPs.

In August 2014, the Nuclear Damage Compensation Facilitation Corporation was reorganized and renamed the Nuclear Damage Compensation and Decommissioning Facilitation Corporation, to include functions such as support for the decommissioning. The management of the reserve fund for decommissioning and related activities needed to carry out decommissioning work was added to its role in May 2017, following an amendment to the act.

3.2.11. Available infrastructure

The waste management strategy can be supported by utilization of available infrastructure and equipment, both on- and off-site. Useful resources can be leveraged from both domestic and international sources. Existing facilities can be leveraged and repurposed, as practicable, to facilitate waste management and support site cleanup. Appropriate off-site facilities can also be included as options to support waste management efforts. Support from unaffected peer nuclear facilities is advisable where available.

During waste cleanup efforts at TMI NPP, an existing fuel pool was initially repurposed for submerged demineralizer system (SDS) storage, taking advantage of the enhanced shielding afforded by the spent fuel pool, with operations being conducted remotely [1]. After the SDS materials were treated and disposed of, the fuel pool was used to store loaded fuel casks that were awaiting shipment, again to take advantage of the enhanced shielding. In the immediate aftermath of the accident, a metal shed that was used for painting was repurposed for solid waste collection and temporary storage. The shed was equipped with a sprinkler system and was located within the main dyke of the plant. The shed was located relatively close to the site boundary and temporary shielding was occasionally needed. It quickly became evident that a more extensive engineered solid waste storage system was needed, and new facilities were constructed.

Similarly, at Fukushima Daiichi, TEPCO repurposed several buildings in the vicinity of the reactors (the Turbine Building, the High Temperature Incinerator Building and the Process Main Building) to support water treatment activities [7]. Many existing water tanks, especially welded tanks for storing concentrated saltwater, were diverted for storing treated contaminated water. In addition, shortly after the accident, tankage and other specialized tools and equipment were obtained from sources worldwide.

The Government, together with corporate suppliers, is conducting remote decontamination activities inside the reactor building and containment vessel. In addition, R&D is ongoing on technologies suitable for extracting fuel debris. Units 5 and 6, which were unaffected by the accident and for which a final decision precluding restart has been made, now serve as full scale test facilities where mockup tests (actual device verification tests) can be conducted.

Regarding reuse of off-site infrastructure, the J-Village stadium (a huge sports complex associated with the national football training centre) was repurposed immediately after the accident as a base for accident response operations. Decontamination of fire engines, military vehicles and helicopters used to cool the spent fuel pool and remove debris was carried out at the repurposed location. The facility also provides housing and break areas used by the Japan Self-Defense Forces, police and other involved organizations. In January 2013, TEPCO established the Fukushima Reconstruction Headquarters at J-Village, supporting more than 4000 employees. Some of the compensation examination work was transferred from the TEPCO head office in Tokyo to the reconstruction headquarters. J-Village was restored to its original function, supporting Japanese football, in April 2019.

During and following the Windscale Pile 1 accident, all the liquid and solid wastes were managed within the existing capacity of the available infrastructure. The fuel and isotopes were mainly discharged, albeit unconventionally, via the existing designed route to the cooling pond, where they were stored and/or processed via the existing fuel cycle route. Filters from the stack were removed, along with other contaminated solid wastes, and disposed of in one of several available solid disposal facilities located on the Windscale Piles site. Water used during the firefighting effort was diverted, where possible, to the pile cooling pond for a short period of decay storage, prior to discharge to sea.

At the time of the Windscale Piles accident, little had been done to prepare for an emergency of this nature. Sir Christopher Hinton (who was responsible for the design and construction of the UK's early civil nuclear programme) later noted that Pile 1 would be "a monument to our ignorance" [60]. As such, the technology deployed during and immediately after the accident was simple and available [60], for example:

- Every steel rod available to the workforce was obtained in order to push the cartridges from the pile;
- Carbon dioxide (taken from stocks held at Calder Hall) was piped into the core with no significant effect;
- Water was then sought as a last resort, with the equipment required to discharge it into the pile having to be improvised. Four hoses from the fire engine were connected to scaffold poles and inserted into the pile.

During the liquidation of the Chernobyl accident and the cleanup campaign, the existing infrastructure was used to the greatest possible extent. The solid waste storage building was utilized for accident waste as well the storage facility initially designed as a waste store for Unit 5, under construction at the time. The later facility was repurposed as radioactive waste disposal site (RWDS) Stage III, also referred to as RWDS Kompleksny in the literature.

In addition, the very large Chornobyl NPP cooling pond was used to host a significant quantity of contaminated material during the initial emergency response and is now in the process of decommissioning.

At present, several facilities at the Chornobyl NPP that are currently undergoing decommissioning are used as temporary storage sites for radioactive materials or wastes. As shown in Fig. 6, the turbine hall of Unit 3 is used to store fragments of the dismantled ventilation stack from the second stage of NPP decommissioning.

3.2.12. Availability of experienced personnel and training needs

Response to an accident is likely to require significantly more resources than normally available within the relevant organizations (including e.g. regulatory agencies), and this may need to be sustained for many months or years. Potentially large numbers of new workers, perhaps with different languages, dialects or cultural backgrounds relative to local workers, may need, for example, basic radiation protection training. It is important that preplanning gives proper consideration to capabilities and knowledge that could be needed for specific roles in support of recovery from a nuclear accident. There will be a need to acquire appropriately qualified individuals, not only to respond to the accident, but also to minimize disruption to other parts of the nuclear industry within that country and to ensure that normal business can resume as quickly as possible. In parallel with recruitment, there will be a need to train new members of staff or to upskill existing staff members to deliver waste management activities. Consideration will be needed of how recovery roles will change over time to allow planning of when to commit resources or release resources for deployment elsewhere.

FIG. 6. Temporary (buffer) storage in Chornobyl NPP Unit 3 turbine hall, with fragments of the dismantled ventilation stack from Units 3 and 4. Courtesy of State Special Enterprise Chornobyl NPP.

The existing staff at a nuclear facility will likely not possess the necessary skills to manage and perform the increased waste management functions after a nuclear accident. Identifying skilled personnel to support the waste management effort will be important for the success of the cleanup programme. There will be a greater reliance on skilled personnel than in normal operations, as many of the activities will not be covered by existing management systems and processes. Strengthening staff resources with both national and international experts in accident and waste management is an important further consideration. The use of international experts can lead to special training needs that address language and cultural differences as well as the requirement to make outside experts fully cognizant of local operational and safety practices. Establishing training programmes will be necessary to ensure worker health and safety. Based on the actual working conditions, which may be very difficult, and the scale of operations, a large work force with shortened work shifts may be a consideration. In some cases, it may be appropriate to contract the waste management function entirely to an experienced entity.

As a rough average, approximately 1000 workers were involved each year in the TMI-2 cleanup. The plant population ranged from a high of almost 1400 in 1985 to a low of approximately 500 in 1989 (at the end of the programme) [2]. Initially, the workforce consisted primarily of personnel from the reactor operator General Public Utilities (GPU) Nuclear. Several volunteers from numerous GPU locations supported the cleanup effort from the outset. A training programme was quickly established to ensure worker safety. The focus of the volunteer workforce was on decontamination activities that freed the more skilled engineers and operators to focus on more challenging problems. The large workforce helped minimize individual worker radiation exposure. Later, contractor personnel were heavily used to support GPU personnel.

To address the uniqueness of the work and challenges associated with an untrained workforce, the following measures were included in the cleanup programme:

(a) Task planning: detailed planning of all work activities to minimize time and include necessary worker safety measures (remote tools, robotics, personnel protective equipment, etc.);
(b) Mockups: rehearsals and work practice assessments to maximize worker effectiveness and efficiency were employed;
(c) Staging areas: areas in low radiation dose zones were utilized to assemble and service equipment, and allow for worker rest breaks, etc.;
(d) Training: an extensive training programme was developed and deployed to ensure that workers were knowledgeable about hazardous conditions and effective work practices.

Similar techniques are being employed at the Fukushima Daiichi NPP and that procedure is described in the decontamination guidelines [61].

3.2.13. Mode of deployment of technologies (fixed versus mobile/transportable)

The scale of the accident and amount/type of waste to be managed can dictate the type of facilities that are deployed. However, the use of centralized facilities versus mobile/transportable facilities can often be application dependent. In some cases, there could be available technologies, including specialized tools and equipment, or units that can be deployed quickly to address an urgent need. The decision making process to select the appropriate technologies or mix of equipment can be influenced by a variety of technical and non-technical factors. Typical selection processes and criteria are described in greater detail in Ref. [51]. In the early stages of an accident or accident recovery, rapid deployment of equipment and services may be required. As such, easily deployed equipment and resources (such as mobile equipment) are often preferred, even though they may not be technically ideal or the most cost effective application for the situation. Consideration can be extended to using mobile/transportable equipment available within other Member States, assuming that the equipment is not contaminated such that transport is ruled out.

The water treatment processes following both the TMI and Fukushima Daiichi accidents relied on centralized treatment facilities that were constructed on-site and operated for extended durations. At the

Fukushima Daiichi NPP, multiple centralized facilities were constructed for water treatment, starting with a caesium removal process, progressing to installation and deployment of the Simplified Active Water Retrieve and Recovery System (SARRY) to augment caesium removal and, ultimately, to the Advanced Liquid Processing System (ALPS) for removal of multiple radionuclides in the contaminated water. Purification of contaminated water in the main trench of Units 2 and 3 has been carried out using a mobile water treatment system. Another mobile water treatment system is used for purification of cooling water from the spent fuel pools. To reduce the level of strontium in reverse osmosis concentrated water, a mobile strontium removal system was installed.

Mobile systems are described in greater detail in Section 9 of this publication and in Ref. [62].

3.2.14. Research and development needs

Due to the complexity of the wastes and hazardous site conditions, existing waste characterization, retrieval and treatment technologies may not be adequate. R&D activities might be required, ranging from development or adaptation of specific use technologies to large scale waste processing systems. A technology assessment exercise using appropriately qualified national and international experts can be highly effective in identifying R&D needs. Using the results of the technology assessment, an R&D plan can be developed to specify R&D needs and define an appropriate technology development plan. In some cases, parallel technology demonstration scenarios can be developed and employed. This approach has been implemented at the Fukushima Daiichi NPP and can help to identify and confirm the most promising technologies in a more expeditious manner.

At the Fukushima Daiichi NPP, an International Collaborative Research Building was established in 2017 to promote R&D from basic research to applied studies, such as improving understanding of the characteristics of radioactive waste. The R&D plan strengthens the core function of bringing together researchers and research institutes and serves as a bridging function between research and decommissioning sites.

3.2.15. Feedback/lessons learned

Previous experiences from nuclear accident cleanups provide valuable information and guidance that warrant consideration as part of the planning processes but are also important for creating and maintaining awareness of relevant old and new industry experiences for the team implementing the waste management programmes and providing guidance in defining requirements. A significant amount of this information is available in the public domain, including Refs [2, 22, 47, 49, 52, 60, 63–66], and specific information is captured in this publication. Publications regarding other industrial accidents (e.g. Bhopal chemical accident, train derailments, oil spills, etc.) can also provide insight into effective cleanup and waste management practices.

The waste management strategy can benefit significantly from employing an active current awareness, feedback or lessons learned programme to support continued technology development and inform implementation operations. Positive and negative practices can be identified and reviewed with personnel on a regular basis. Establishment of a continuous improvement mindset can be an effective means of driving excellence in safety, cost effectiveness and conduct of operations.

A concerted effort was made after the TMI accident to document the results of the accident, including the cleanup and waste management effort. These publications are available in the open literature and are referenced within this publication. For example, a series of reports summarizing several aspects of the cleanup and waste management effort was prepared. These reports, known as GEND reports, were developed under the direction of General Public Utilities (the TMI NPP facility operator), Electric Power Research Institute, US Nuclear Regulatory Commission and US Department of Energy [66].

A large number of reports and scientific articles were prepared after the Chernobyl accident, describing lessons learned. The literature includes IAEA publications (e.g. INSAG Series No. 1 [64] and INSAG Series No. 7 [65]), as well as numerous articles in Ukrainian and Russian language scientific journals, and articles in international scientific journals. As the cleanup progress evolves at the Fukushima Daiichi NPP, reports and scientific publications have been prepared by numerous agencies and countries (some of these are identified in Appendix IV).

3.2.16. Interactions with other programmes

The waste management strategy and related requirements development can take advantage of interfaces with other related entities, including nuclear materials accounting and safeguards programmes, civil emergency management authorities, local and regional ministries, national governmental bodies, neighbouring states and international organizations. Maintaining lines of communication, as appropriate, will be important to gain acceptance of the waste management strategy.

Logistical, resource, technological and expert support may be available at other similar or related national facilities, as well as in other Member States. Consultation and interaction with these entities, the IAEA and other international organizations can augment the waste management team's capabilities while not being allowed to distract from implementation of the most urgent waste management activities. Again, the importance of preplanning that includes interactions with relevant agencies is emphasized. In the event of an accident, the organization responsible for waste management will need to identify gaps in national resources in order to best focus international support on those areas where value can be added.

It was quickly realized after the TMI accident that coordinated involvement of both commercial and governmental entities was needed to ensure successful environmental cleanup. A unique consortium was established in 1980 to leverage the expertise of national organizations within the United States of America. The GEND agreement, discussed above, resulted in coordinated efforts by personnel at GPU, the Electric Power Research Institute (EPRI), the US NRC, and the US DOE [2, 66].

For the Fukushima Daiichi accident, the IRID was established in 2013 to promote R&D in the area of nuclear decommissioning under an integrated management system. IRID efforts are focused on the most pressing issue, the decommissioning of the Fukushima Daiichi NPP, with a view to strengthening the foundations of decommissioning technologies. The members of IRID are the Japan Atomic Energy Agency, the National Institute of Advanced Industrial Science and Technology, 4 manufacturers and 12 electric utilities.

3.2.17. Communications needs and public involvement

Environmental cleanup and associated waste management after a nuclear accident will be a complex and long term challenge with significant interest in the restoration of the site and surrounding areas and the future of local communities from a variety of stakeholders, including government officials, community leaders, the nuclear industry and the public. One of the pillars of the environmental cleanup and waste management strategy is to implement processes that provide for effective stakeholder engagement. Stakeholder engagement is also a key component of community revitalization, by bringing together concerned stakeholders to establish a vision for the future. Requirements derived from stakeholder engagement efforts are an essential component of overall requirements management.

Members of the public are likely to seek information from a range of sources, including facility operators, regulators and local/national governmental organizations and, in some cases, from less well or potentially ill informed unofficial sources. It is important that there is a degree of coordination and liaison between organizations to ensure the coherence and accuracy of messaging. The roles and responsibilities of these organizations in the area of communication need to be clarified early, as there will be a need for proactive and reactive communications. The various organizations may emphasize different issues or topics, so proper coordination is needed to avoid factual differences and minimize misunderstandings. Consistent delivery of fact based information is essential in building and maintaining public trust. The IAEA developed IAEA Safety Standards Series No. GSG-14, Arrangements for Public Communication in Preparedness and Response for a Nuclear or Radiological Emergency, which describes the infrastructure and processes needed to provide useful, timely, truthful, consistent, clear and appropriate information to the public in the event of a nuclear or radiological emergency [67].

It was acknowledged that there was a breakdown in public communication shortly after the accident at TMI NPP, causing strained relationships with the public. Further, TMI NPP became a focal point for anti-nuclear groups and politicians [2]. This continued to be an issue throughout the cleanup campaign, as management attention was often diverted to address public concerns. A concerted public relations effort, including frequent release of information and demonstration of programme transparency, would have alleviated some of these problems.

To help people understand the plans and efforts underway to decommission the Fukushima Daiichi NPP, in 2013 TEPCO created a Risk Communicator position and established the Social Communication Office [68]. These were created in recognition that the corporate culture was neither supportive of risk recognition nor capable of communicating risk openly to the public. A mindset of infallibility had existed within the company and had been projected outward into society at large. The Social Communication Office, along with Risk Communicators, were established to promote fact based communication focused on establishing trust. They are tasked with addressing perception gaps, promoting public relations and communicating potential risks.

The experience of the MOE was that there have been serious difficulties in communication with stakeholders, such as residents being obliged to evacuate or those in areas requiring decontamination or located near waste storage sites. There were some misunderstandings between the different sections of central government and the MOE, or with TEPCO. However, through proactive measures it has been possible to obtain positive and active cooperation with significant effort from academics and other specialists.

Experience at the Fukushima Daiichi NPP has clearly demonstrated that there may be a need for numerous and repeated individual and group meetings in order to engage with the public's emotional and perceptional reaction. These interactions are important in harmonizing public understanding and gaining acceptance for the execution of decontamination work. It has only been since the successful implementation of related efforts, over a roughly three year period, that significant progress has been made in executing the decontamination work. In fact, the schedule for execution of these works was driven to a significant extent by these consensus building efforts within the community as opposed to simply overcoming technical problems. The performed activities at the decontamination locations, along with the relevant safety explanations, were summarized and published by the MOE in 2018 [69].

Problems in risk communication and problem solving techniques are being actively studied by researchers, both in Japan and internationally (e.g. an IAEA publication [70] and the NEA 2020 Draft [71]). However, it is noted that risk communication techniques will not be universally applicable and will always need to reflect the specifics of the national character and policy decisions in effect at the time. It is noted that the nature of communication required can be very different, depending on the societal needs in the region that has suffered a nuclear accident.

Regardless of the differences, however, there will be some areas of commonality, such as:

(a) Providing actual locations such as a local information centre (rather than only web based contact points) where members of the public can obtain information and learn about basic issues such that they can judge the situation for themselves. These facilities are effective for the mutual exchange of information and communication relating to the challenges being faced. Relying on mass communication channels, such as the Internet, is not sufficient, especially after natural disasters, when infrastructure might be degraded.
(b) Having small group meetings to provide explanations, with size limited to a manageable number, such as 20 people.
(c) Efforts to create a relationship of mutual trust through use of presenters at local meetings whom the local population trust and respect.

4. PRE- AND POST-ACCIDENT PLANNING

Key lessons learned:

- Preplanning that covers a wide range of potential accident scenarios and addresses organizational roles and interactions will help to facilitate understanding of the available waste management options and to prepare for the initial rapid response to an accident.
- Review of past accidents and how they were handled (lessons learned — both good and bad) will inform waste management strategies and plans.
- The extent of preplanning required is dependent on the nature and scale of the nuclear facilities and credible accident scenarios in the Member State and surrounding Member States.
- Preplanning allows identification of essential infrastructure that is, or can be, available for waste management activities.
- Predisposal: identify methods and procedures for conducting waste retrieval, processing, storage and disposal operations, and identify where repurposing or adapting existing facilities to new accident waste management facilities would be feasible, especially in the early stages of an event, when facilities are required quickly.
- Disposal: identify disposal options that might be available at both a national and a local scale, including the potential for siting and constructing new disposal facilities.
- An overview of potential disposal options, repository design concepts, locations, capacities and disposal strategies will provide the basis for conceptual waste acceptance criteria. This will have a major impact on immediate decisions on segregating and treating wastes, from the outset of post-accident waste management activities.
- The way the organizations react to an accident event and its subsequent cleanup will have a large influence on the quantities and types of radioactive wastes that need to be managed.
- The use of independent experts to review and advise on plans and technologies can be helpful, especially in evaluating proposed solutions.

While there is substantial guidance available for developing waste management policies and strategies for normal situations [72], these rarely take into account the unique needs of post-accident situations, which are described in this section.

Historical experience with accidents (see the appendices) involved extraordinary, unforeseen conditions, such as the release of core material from confinement and extraordinary volumes of very low to extremely high activity waste to the environment that significantly exceeded the conditions anticipated in the safety analysis and licensing processes for the facilities. The resultant waste was correspondingly challenging to retrieve, collect and store, particularly with respect to the attendant impacts on worker and public doses, on the environment, and on national, regional, local and facility owner economics.

Preplanning for managing the response to potential accidents and the radioactive materials and wastes that could result from them is likely to require the involvement of several organizations in a Member State and the commitment of resources to comprehensive joint exercises. Preplanning can identify a wide range of scenarios for analysis that are relevant to the specific nuclear facilities in a Member State, including consideration of exercises that include relevant national (and international) organizations that could become involved in accident response. Advanced preplanning exercises might also consider possible accidents in nuclear facilities in neighbouring countries. Identifying and understanding the range of credible accident scenarios will allow scoping of potential source terms, potential release pathways and radioactive waste types, along with the inventories that may be involved and how these could evolve with time after an accident. Information on arrangements for radioactive waste management during and after an emergency can be found in IAEA Safety Standards Series No. GSG-11, Arrangements for the Termination of a Nuclear or Radiological Emergency [73].

Management of radioactive wastes is, of course, only one aspect of preplanning responses to potential nuclear accidents. Much of the emphasis in preplanning exercises will be on managing releases of radioactivity and reducing immediate population and worker exposures. Common aspects of all accident preplanning exercises are that they:

(a) Involve and test the roles and interactions of the national and regional nuclear safety, radiological protection and emergency planning and response agencies, along with the owners and operators of all relevant nuclear facilities;
(b) Are based on accident scenario identification for specific facilities that can be used to determine the types of accident, their possible event sequences, the materials involved and the areas that could be affected;
(c) Consider stakeholder, societal and resource constraints on the extent of practical preparedness that can be achievable for each scenario and facility and how planning implementation can be prioritized;
(d) Define facility specific strategies and plans for a range of scenarios, which can be deployed rapidly and effectively by the agencies involved in the event of an accident occurring.

Prescreening and identification of potentially applicable technologies is a useful planning exercise to be conducted when developing response plans for a potential nuclear accident. Many Member States already engage in such activities as part of their emergency preparedness programmes [74]. An example is shown in Fig. 7.

Those parts of strat`egic preplanning exercises that deal specifically with waste arisings and establishing a waste management strategy and system might be expected to:

(a) Develop qualitative and quantitative estimates of the possible event consequences, including radioactive material released and radioactive waste evolved during each phase of the accident, in order to bracket or range the parameters needed for preplanning: for example, form, radionuclide

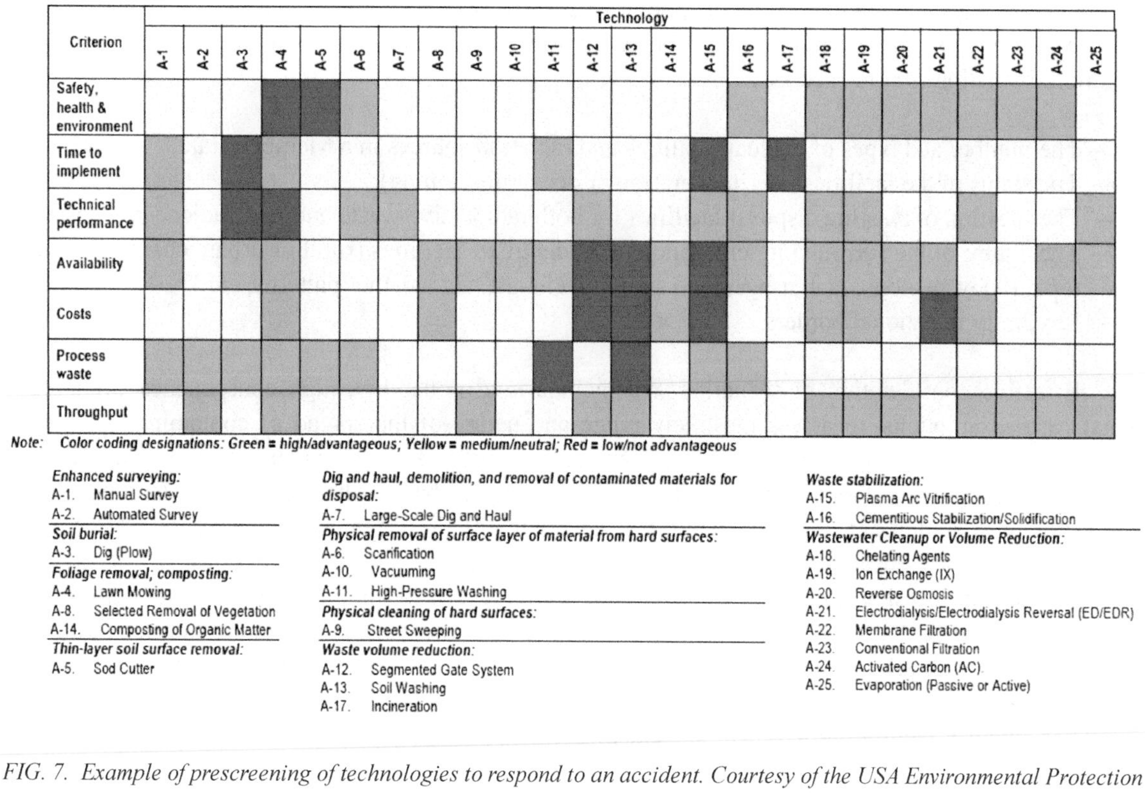

Criterion	Technology																								
	A-1	A-2	A-3	A-4	A-5	A-6	A-7	A-8	A-9	A-10	A-11	A-12	A-13	A-14	A-15	A-16	A-17	A-18	A-19	A-20	A-21	A-22	A-23	A-24	A-25
Safety, health & environment																									
Time to implement																									
Technical performance																									
Availability																									
Costs																									
Process waste																									
Throughput																									

Note: *Color coding designations: Green = high/advantageous; Yellow = medium/neutral; Red = low/not advantageous*

Enhanced surveying:
A-1. Manual Survey
A-2. Automated Survey

Soil burial:
A-3. Dig (Plow)

Foliage removal; composting:
A-4. Lawn Mowing
A-8. Selected Removal of Vegetation
A-14. Composting of Organic Matter

Thin-layer soil surface removal:
A-5. Sod Cutter

Dig and haul, demolition, and removal of contaminated materials for disposal:
A-7. Large-Scale Dig and Haul

Physical removal of surface layer of material from hard surfaces:
A-6. Scarification
A-10. Vacuuming
A-11. High-Pressure Washing

Physical cleaning of hard surfaces:
A-9. Street Sweeping

Waste volume reduction:
A-12. Segmented Gate System
A-13. Soil Washing
A-17. Incineration

Waste stabilization:
A-15. Plasma Arc Vitrification
A-16. Cementitious Stabilization/Solidification

Wastewater Cleanup or Volume Reduction:
A-18. Chelating Agents
A-19. Ion Exchange (IX)
A-20. Reverse Osmosis
A-21. Electrodialysis/Electrodialysis Reversal (ED/EDR)
A-22. Membrane Filtration
A-23. Conventional Filtration
A-24. Activated Carbon (AC)
A-25. Evaporation (Passive or Active)

FIG. 7. Example of prescreening of technologies to respond to an accident. Courtesy of the USA Environmental Protection Agency [74].

content, activity ranges, exposure mechanisms and radiation doses, volumes, locations, effects on plant functioning, ability to retrieve, etc.;
(b) Identify potential waste management strategies and systems and the requirements that will drive decisions on the selection, design and operation of each component of the waste management system, using a system engineering approach;
(c) Identify potential components of a facility specific waste management system for material collection storage and disposal: for example, retrieval and excavation equipment; collection and storage locations both near and away from the facility; capacity and proximity of disposal capability; potential repurposing of facilities; transportation and related logistic capability; etc.;
(d) Determine the financial, personnel and equipment resources required and available for each scenario and develop corresponding plans for managing the anticipated wastes;
(e) Identify gaps in capabilities and facilities that would lead to failure to meet key requirements and ensure that these are filled.

The waste management strategy and implementation plans need to consider the progressive phases of an accident event. In the initial (or emergency) phase, decisions need to be made quickly and the priorities are controlling the situation and mitigating consequences rather than optimizing waste management [45]. As the accident moves into the recovery and cleanup phases, there is more time for planning, optimizing and executing a waste management strategy. The predisposal strategy and plans need to consider, and be integrated with, the disposal plan and the cleanup and remediation strategies. Predisposal waste management activities will likely need to be adaptable to respond to inevitable changes in these other plans.

In the event of an accident, detailed waste management planning can only take place after the initial emergency has been stabilized, when the full extent of the situation is better understood. National strategic preplanning, when adequately performed, will provide a firm basis for more detailed tactical planning and will highlight critical questions to be considered in the plan. The topics addressed in the sub-sections of pre- and post-event planning later in this section are provided solely as examples. The degree of actual preplanning employed will depend on the policies and organizations in individual Member States.

As discussed abov`e, preplanning is based on scenario exercises that take account of the national situation, including factors such as:

— The number and types of nuclear facilities and radiation sources in a Member State;
— The status of the facilities (i.e. in operation or decommissioning);
— The location of existing disposal facilities for both radioactive waste and non-radioactive wastes;
— The nature of the surrounding environment at identified facilities (rural or urban, coastal or inland, topography, geology, land usage, land stability, climate and weather patterns, etc.);
— Proximity to national borders.

Effective consideration of scenarios will include low probability, high consequence events, and worst case assumptions, to assess the likely range and nature of the spread of contamination and the wastes generated. The Fukushima Daiichi accident showed that it is not reasonable to screen out scenarios on the basis of poorly understood probabilities or existing design bases. The preplanning planning process can, with consideration of the above event types, ensure that the waste management response is adequate to handle the consequences of the scenarios considered.

In addition to using scenarios, consideration can be given to undertaking an exercise to test the efficiency of the plan and its implementation. Revisions to the plan can then be made, if required, afterwards. As with all such activities, it is important to update the plan at regular intervals to reflect changing circumstances.

For nuclear facilities, there will be existing emergency response plans. However, these plans typically do not address waste management. Consideration can be given to expanding these emergency response plans to address waste management during the early urgent phase to provide appropriate actions before the national response is initiated.

For countries that do not have nuclear facilities, a smaller, proportionate preplanning process for waste management can be undertaken. This will need to address scenarios based on accidents involving radioisotopes and radiation sources, and possible malicious actions.

An organizational structure can be identified from the preplanning exercises that can be put in place immediately after an accident occurs, to implement and adapt the waste management strategy. A thorough understanding of potential accident consequence ranges and the pertinent regional and national waste management policies and regulatory requirements is needed in advance of an accident to ensure that accident and longer term waste management responses are responsive to the event conditions and are consistent with applicable regulations and policies.

An example of preplanning is provided by the CODIRPA[2] group in France. An interministerial directive in 2005 on the action of public authorities in the event of a radiological emergency tasked the Nuclear Safety Authority with establishing the framework and defining, preparing and implementing the measures necessary to respond to situations following a nuclear accident. The Nuclear Safety Authority set up a steering committee (CODIRPA) that was responsible for developing the approach and methodology to be used. This committee brought together various stakeholders, including the main ministerial departments concerned, expert bodies, associations, elected officials and operators of nuclear installations, and sought international expertise.

A further example of a methodology to conduct preplanning is provided in Appendix 1 of Ref. [21].

4.1. PREPLANNING FOR DISPOSAL

In preplanning for the disposal of accident wastes, it is important to understand that the availability of technical, financial and regulatory resources can affect the timing of the different steps to be taken. For example:

(a) Rapid disposal: a Member State with the necessary resources, experience and infrastructure may choose to adopt a strategy aimed at relatively rapid disposal. In this situation, preplanning would place greater emphasis on the identification of existing facilities that could safely be adapted or expanded, using one or more additional disposal facilities or disposal concepts not previously employed, and/or identification of potentially suitable locations for siting of new facilities, with application of existing licensing procedures to the extent possible. Rapid disposal is appropriate where adequate site information is available and locations have been assessed as suitable in preplanning exercises — this will avoid the situation where, in the event of an accident occurring, future problems are created by developing uncharacterized and unsuitable disposal sites.
(b) Deferred disposal: until a Member State is confident that it has satisfied all its internal social, political and technical issues related to siting, construction and licensing a disposal facility, it may prefer a strategy that defers disposal. In such a case, preplanning would place greater emphasis on interim management sites or facilities, whether as limited duration staging areas or for longer term storage. Preplanning for interim management would ideally address topics such as location criteria, licensing, construction standards and operational practices.
(c) Hybrid strategy: it is also possible to define a hybrid strategy, in the sense that there may be some anticipated waste types that could be relatively easily disposed of in existing facilities, while a longer term approach may be adopted for more challenging waste types or to accommodate very large volumes.

[2] CODIRPA: COmité DIRecteur pour la gestion de la phase Post Accidentelle d'un accident nucléaire ou d'une situation d'urgence radiologique.

4.1.1. Evaluation of existing disposal capacity

In some countries, existing facilities may be sufficient to manage the immediate needs for most or all of the waste arising from even large accidents. In others, disposal facilities might not exist, or capacity might be insufficient to accommodate wastes from reasonably possible accident scenarios.

During the preplanning phase, it is useful to consider regulated disposal facilities for multiple waste types, not only those accepting radioactive waste. In some Member States, non-radioactive waste disposal facilities could possibly be upgraded or expanded to allow disposal of specified types of high volume, low activity waste generated from an emergency consistent with a properly developed safety case.

A robust planning process would include identification and evaluation of the following questions regarding existing disposal facilities and waste handling capacity:

(a) Land area occupied and location in relation to existing NPPs, military installations or other potential accident locations, waste treatment and conditioning facilities, and rail and highway transportation routes;
(b) Waste types allowed, total permitted disposal capacity in terms of volume and activity, additional planned disposal capacity, historic usage rates and facility expansion potential;
(c) Existing laws, regulations, permit conditions, community agreements or social equity factors that might allow, prevent or limit facility use or expansion for waste generated in an emergency;
(d) Environmental conditions, buffer zone or infrastructure limitations, future land use plans, or other factors that could prevent or limit facility use or expansion for accident wastes.

4.1.2. Evaluation of potential new disposal facility locations and technologies

IAEA Safety Standards provide siting requirements that define what constitutes a technically sound disposal site that will protect human health and the environment. Individual Member States have varying approaches to economic, social and cultural siting criteria, as well as environmental criteria that have no direct bearing on the safety case. A siting process that effectively combines all types of siting criteria enhances prospects for success. However, the weighting placed on the criteria involved in siting a disposal facility, particularly in a zone seriously contaminated by a nuclear accident, will be different from that required for the siting of a disposal facility in otherwise radiologically unrestricted areas under non-emergency conditions.

If existing disposal facilities have limited capacity or poor expansion potential, it will be useful to identify general areas possessing favourable characteristics with respect to hydrogeology, geology, surface water drainage and seismic activity in advance of an emergency. This scoping offers several advantages:

(a) Early identification of objective siting criteria, as well as opportunities to collect and discuss information concerning these criteria with relevant stakeholders in the absence of actual emergency conditions;
(b) Shortened time frames for location and authorization of new sites that might be needed after an actual accident;
(c) Enhancement of public confidence through establishment of a group of informed public stakeholders who participate in the early scoping process, which can facilitate communications that may enhance public confidence in decisions made after the event of an actual accident.

This scoping can use available data or new data if funding is available. However, since the selection of disposal sites can be a highly sensitive issue, it might only be appropriate in preplanning to conduct relatively broad studies of options in the area around the nuclear facility being evaluated. The final selection of the disposal site(s) and technologies would take place if an accident occurred.

In considering potential disposal concepts that can be preapproved, experience with facilities that have defined design requirements might be easier to evaluate than concepts that would require extensive site specific evaluations.

An example of preplanning consideration of potential disposal facilities is provided by the work of CODIRPA in France (discussed earlier in this section). The CODIRPA preplanning deliberations did not initially identify the Morvilliers CIRES facility (see Appendix VI) as an appropriate prototype for new facilities for the management of large volumes of radiologically contaminated wastes that could result from an accident, as it was not certain that geological conditions in the region of an accident would be similar to and meet the same site suitability criteria applicable to Morvilliers. CODIRPA is assessing whether the principles used for the Morvilliers facility (regulation for hazardous waste disposal) can be used for the development of a dedicated facility that could also receive very low levels of radiological contamination, with a local construction (reconstitution) of the natural barrier with imported material such as clay, as is performed for hazardous waste landfills.

Similarly, specific design and engineering requirements are identified for chemically hazardous waste landfills in the USA that have been determined to provide the necessary level of containment across a range of geological and meteorological settings. These requirements (Code of Federal Regulations: 40 CFR Part 264) include elements such as the design and materials to be employed for liners and leachate collection systems, stability requirements and final cover designs. Coupled with basic siting criteria (e.g. avoidance of floodplains or seismically active areas), such facilities have operated successfully in both humid environments with relatively shallow ground water, and arid environments, with very deep ground water. Many facilities have also been used to dispose of waste that contains relatively low concentrations of radionuclides, either by exemption or originating from activities such as oil and gas production. Versatility and adaptability to different environments are desirable for preplanning purposes.

In the USA, regulatory authorities issue standard disposal facility permit application content requirements and standard regulatory review plans for LLW disposal. The availability of these standard forms shortens the time required to assemble and later review permit applications for new disposal facilities.

4.1.3. Averaging, dilution and blending approaches

Discussion of waste volumes in the context of a nuclear or radiological emergency would be incomplete without considering the potential effects of intended and unintended dilution on total disposal volumes. For the purposes of this publication, intended dilution refers to controlled mixing or blending of wastes exhibiting higher and lower activity levels, or uncontaminated material, to reduce overall activity levels and allow authorized disposal at a licensed facility. These approaches are typically developed by regulators based on appropriate safety case analyses.

Unauthorized dilution to subvert permitted disposal criteria is not an internationally accepted practice. Authorized dilution may be part of a graded approach to disposal, offering cost savings through appropriate use of multiple disposal facilities. For the purposes of this publication, unintended dilution refers to actions taken (or not taken) in the aftermath of an emergency that led to increasing the contamination levels (and likely the waste volumes) of previously uncontaminated or slightly contaminated material. Such unintended volumes can have a major impact on future disposal capacity needs and related resource requirements.

Placement of accident waste in unlined trenches or other poorly contained conditions can result in significant contamination of surrounding soils and groundwater and surface water bodies through migration of radionuclides out of burial units. As shown by the Chornobyl NPP experience discussed herein and the various legacy sites described in Appendix VI, unintended contamination is particularly likely in areas with high precipitation and permeable soils. Adverse impacts tend to increase for long lived heterogeneous wastes. Over time, the volume of contaminated surrounding soils can therefore increase. In the case of an accident, temporary storage in properly contained conditions may be warranted, pending the availability of properly designed disposal facilities, which may take several years to decades to become available.

Ploughing under contaminated soil to reduce ground surface contamination levels dilutes radionuclide concentrations within a larger depth of contaminated soil. A later decision to excavate and dispose of such contaminated soils could result in large waste volumes requiring disposal. While ploughing under may be undertaken with no intention of future excavation, changing land use plans for agricultural purposes, population growth or other reasons could lead to a reversal of this initial decision [75].

4.1.4. Stakeholder involvement in accident waste disposal

Informing and involving a wide range of stakeholders, especially those affected by an accident, is critically important to nuclear accident response planning and strategy development, including waste management and disposal aspects. The main objectives of an involvement and communication programme are to build public trust, encourage broad participation and maximize acceptance of proposed disposal solutions, while minimizing negative public perceptions based on unanswered questions or misinformation. Providing information and seeking input on emergency waste disposal can lead to multiple benefits, for example:

- The disposal plans and strategies developed can better reflect local needs and preferences, increasing the likelihood of a broad consensus in support of the plans and strategies that are ultimately put in place;
- Sensitive subjects can be explored in a non-emergency atmosphere, encouraging thoughtful dialogue, diverse viewpoints and fact checking from knowledgeable sources;
- Communication mechanisms and trained personnel can be established so that in the event of an actual accident information sharing can be implemented through an already established network;
- Credible third party information sources can be more efficiently made available to the media and the public post-accident to provide rapid, accurate information and dispel misinformation.

The following questions about possible disposal options, facilities and locations can help initiate these dialogues:

- Is expansion of capacity at existing radioactive waste disposal facilities practical? If so, to what extent is such expansion desirable or appropriate? Could multiple waste management facilities be utilized to spread the impact?
- If new disposal facilities might be needed, what process would be used to locate suitable sites? Ought generally favourable locations to be preidentified?
- To what extent would siting of new facilities negatively impact on current or intended land use plans in affected communities?
- Ought economic incentives to be provided to communities accepting the major expansion of existing waste disposal facilities or the siting of new facilities? If so, how ought that process to occur?
- Are changes in legislation, safety standards or regulations needed to expedite new disposal facility authorization or increase operational flexibility at existing facilities in case of an accident? Ought regulatory exceptions consistent with safety and relative societal risks to be evaluated?

The processes employed and stakeholders involved in addressing these questions will vary according to individual Member State legislation and policy. The IAEA has published a number of publications on stakeholder involvement, including inclusive approaches to selecting new disposal sites and technologies, incentive programmes and other topics (e.g. Refs [70, 76]).

4.1.5. Other technical issues in disposal preplanning

Many technical issues need to be considered when planning how to implement the disposal of accident wastes. A gap analysis can be an effective support to preplanning. IAEA [21] suggests:

(a) Performing a gap analysis to evaluate the capability and capacity of existing disposal infrastructure, and also for other tools, methods, etc. that will be needed to implement Government disposal policy in the event of an accident (based on the scenarios considered).
(b) Proposing new work needed to address any identified gaps; this may include research and development to improve existing disposal capability or develop new approaches.
(c) Estimating possible waste inventories for postulated emergencies that reflect the national situation (taking account of the numbers, types and locations of nuclear facilities and users of radioisotopes and radiation sources).
(d) Estimating inventory needs to include likely spread (plume) of contamination, mobility of contaminants in the environment, volume and characteristics of contaminated materials (e.g. soils, trees, buildings, etc.).
(e) Considering the development of modular and scalable designs for disposal of accident wastes that may enable rapid licensing and be implemented quickly.
(f) Evaluating the possibility for the transfer of existing licensed disposal facility designs to allow the rapid implementation and licensing of new facilities for accident wastes — noting that if the waste types, facility designs and locations of proposed new facilities are similar to the respective existing facilities, then their safety performance would also be expected to be similar.

4.2. POST-ACCIDENT WASTE MANAGEMENT PLANNING

Waste management planning after an accident has been stabilized is conducted in a systematic manner with the primary goal of protection of human health and wellbeing, as well as minimizing the amount of radioactive waste that needs to be managed over the long term.

Response to the accident will occur in phases. In conducting planning activities it is important that the various phases of the response effort are well considered, including transitions between phases. IAEA Safety Standards Series No. GSG-11 [73] provides information on arrangements for the transition phase from emergency response to termination of the nuclear or radiological emergency, including radioactive waste management activities. Identified end points can facilitate the transition to subsequent phases and provide tangible measures of the success of the response effort. For example, at TMI NPP, as the cleanup progressed, four operational phases became evident [2]:

(a) Stabilizing the plant (1979–1980);
(b) Waste Management (1980–1983);
(c) Decontamination (1981–1985);
(d) Defuelling (1984–1990).

The phases overlapped and activities within these phases generally did not have discrete start and stop dates. Within the first 15 months after the accident, cold shutdown of the reactor was achieved, fission gases were purged from containment, a systematic approach to control contamination was employed and processes to treat contaminated water were put in place. The waste management phase comprised processing of contaminated water, retrieval of wastes, immobilization of wastes and finally storage and shipment of radioactive wastes. The decontamination phase was primarily directed towards reducing worker exposure and providing access to facilitate cleanup and eventual fuel removal. The defuelling stage was complex and initially involved developing strategies for fuel debris characterization and removal approaches.

This section now looks in more detail at the two examples of post-accident planning provided by experiences at the Chornobyl and Fukushima Daiichi NPPs, which are described in Sections 4.2.1 and 4.2.2.

4.2.1. Example 1: recovery operations planning at the Chornobyl Unit 4 site

A thorough analysis was required for detailed planning of actions and measures to respond to the complex and severe post-accident conditions at Chornobyl. Planning, documented in Chornobyl Unit 4: Short and Long Term Measures[3] [1], was initiated in early 1996 to analyse options and provide detailed recommendations to achieve ecologically safe conditions. The analysis was conducted by an international team of experts, which, together with Ukrainian experts, ascertained the measures needed and the approaches that could be taken to achieve them. A final report summarizing the findings was issued in November 1996. The report included a recommended course of action to be taken as a primary output. Based on the findings of the report, Ukraine, G7/8 countries and the European Commission directed the team to prepare the Shelter Implementation Plan (SIP) [1]. The SIP was completed in June 1997 and approved at the 23rd G8 Summit in Denver, Colorado, in June 1997. The development activities, decision methodology and resulting plans are summarized below.

4.2.1.1. Priorities and decisions for conversion of the Chornobyl Unit 4 site

The international team adopted a deficit analysis approach to establish the objectives for the transition of Chornobyl Unit 4 to an ecological safe condition. The main objective of the expert team is protection of the public, workers and the environment, which requires safe storage of all radioactive and nuclear materials. After extensive dialogue, a common understanding developed that Chornobyl Unit 4 could not be converted into a safe permanent storage facility for the nuclear materials. Consequently, removal of the inventory was determined as the ultimate, long term goal, following a stepwise graded approach to assuring safety, as illustrated in Fig. 8. The implementation time frame for disposal of the radioactive inventory remains open, pending future decisions.

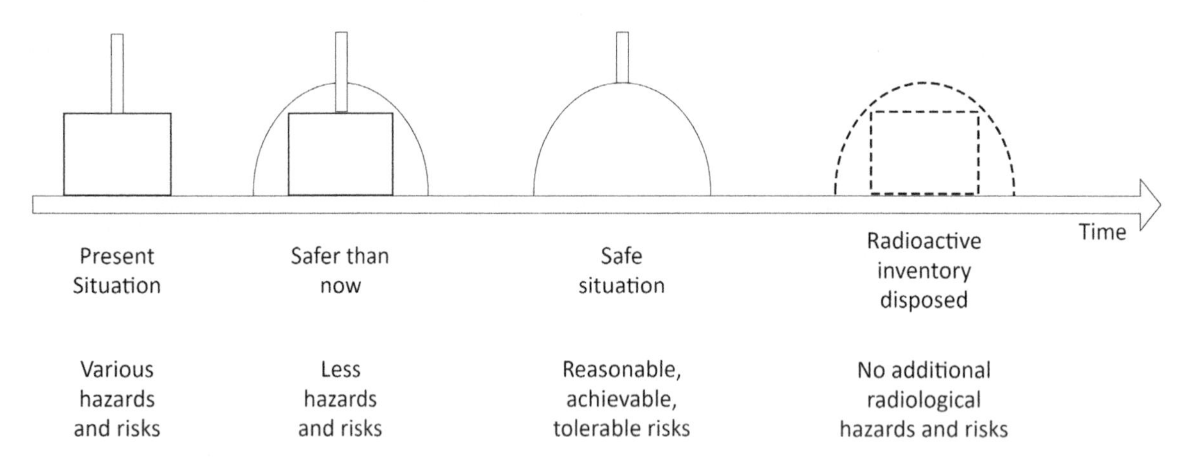

FIG. 8. Staged approach to convert the Chornobyl Unit 4 site into an environmentally safe condition. Adapted from Ref. [1].

[3] 'Chornobyl Unit 4' refers to Unit 4 of the Chornobyl NPP that was damaged in the April 1986 nuclear accident. Chornobyl Unit 4 is often referred to as the 'Shelter' or 'Shelter Object', the Russian translation is 'Ukritiye' and the Ukrainian translation is 'Ukryttja'; sometimes the term 'Sarcophagus' is also used.

In the next step, the experts concluded that it would be necessary to develop and implement measures to systematically address the following specific safety issues:

(a) Source related risks;
(b) Nuclear safety: fuel/fuel containing material (FCM) (fissile material);
(c) Radiological safety: radioactive materials, including core debris;
(d) Other hazardous materials: burnable, explosive, toxic materials;
(e) Operational risks;
(f) Dangerous work functions and activities, industrial safety;
(g) Dust, radiation;
(h) Access ways, industrial safety;
(i) Accidental risks;
(j) Unit 4 — Shelter collapse;
(k) Fire;
(l) Emergency systems and preparedness.

The safety issues were arranged into action categories and were further analysed in terms of urgency, feasibility and the possibility of early implementation. This activity resulted in a list of strategic measures that were urgent, necessary and desirable. The details of the process and associated issues can be found in the related report [1].

Once the objectives and strategic measures were identified and prioritized, a safety based decision tree related to risk mitigation was prepared. By abstraction, three top decisions were to be implemented: short term risks were to be addressed with urgent measures in a short time frame with near term funding as a top priority (Decision No. 1). In the next step, a decision was required on whether, following internationally accepted standards, the inventory could remain at its current location (Decision No. 2). Based on a common understanding, it was concluded that the site cannot be converted into a repository that would meet these standards. The next decision required (Decision No. 3) was on the timing of inventory removal and transfer to a repository. Decision No. 3 is pending at the time of writing this publication.

4.2.1.2. Scenario analysis and development of a recommended course of action

Various scenarios were proposed by the experts and expert organizations supporting the project. These were analysed in detail, including their feasibility, implementation risk, cost and timing. After analysis of each proposed scenario and a subsequent comparative analysis of their pros and cons, a Recommended Course of Actions was developed to provide an organized basis upon which both the short and long term risks could be addressed. A specific scenario study, the Alliance Study, resulted. This study proposed comprehensive measures for creating a new, long lived, arch type, double walled, concrete containment shelter with supporting facilities to confine and contain the damaged reactor unit safely for a period of at least 100 years. Other scenarios considered were, for example, the structural stabilization of the existing Shelter Object, the containment of the damaged reactor in a heavy monolithic concrete structure, or the creation of light confinement structures. The most favourable solution as ultimately implemented is the NSC, which was successfully moved into its final position in 2016. For reference, the tasks associated with each phase are excerpted below [78]:

"Phase 1: Immediate actions and other Short Term Measures

- — Task 1.1. Reduce collapse accident probability by structural stabilisation (to be coordinated with Tasks 2.1. and 2.2.)
- — Task 1.2. Reduce collapse accident consequences (to be coordinated with Tasks 2.1. and 2.2.)
- — Task 1.3. Increase nuclear safety by criticality control and contained water management (as a part of an integrated monitoring system)

— Task 1.4. Increase workers' and industrial safety (appropriate monitoring and safety equipment are necessary)
— "Phase 2: Preparation of the long term conversion into an environmentally safe site
— Task 2.1. Provide safer accesses by shielding and stabilising as far as possible with cemented material fill and dust fixatives (to be scheduled in parallel with implementation of Phase 1)
— Task 2.2. Provide a confinement and remove unstable upper parts. Alternatively the confinement ought to be designed to take a collapse
— Task 2.3. Develop a removal strategy and optionally implement an early partial removal
— "Phase 3: Conversion into an environmentally safe site
— Task 3.1. Convert the site into a safe structure (to be scheduled in parallel with implementation of Phase 1)
— Task 3.2. Control and maintain the safe structure until removal (to be scheduled in parallel with implementation of Phase 1)
— Task 3.3. Removal of remaining inventory when appropriate and necessary."

The Recommended Course of Actions was developed with the intention of defining a framework under which the accident site can be transitioned to an environmentally safe condition. The resulting framework provides a stepwise but flexible approach to taking decisions that is also supportive of future optimization measures. The Recommended Course of Actions was developed between May 1996 and December 1997.

4.2.1.3. Shelter Implementation Plan: current status and results achieved

Addressing the probability and consequences of a shelter collapse, Tasks 2.3 and 3.1–3.3 dealt with many uncertainties associated with implementation of Tasks 1.1–2.2. The consensus was to proceed with the implementation of Tasks 1.1–2.2, with a focus on providing a considerable improvement in safety and a better basis for subsequent decisions needed under Tasks 2.3–3.3.

The Government of Ukraine, the G7 countries and the European Union decided in early 1997 to implement Phase 1 and Tasks 2.1 and 2.2 of Phase 2 and to prepare in parallel the decision base for the subsequent steps (Task 2.3 and the Phase 3 tasks). The SIP developed by the international expert team is the implementation plan covering Phase 1 and Tasks 2.1 and 2.2 of Phase 2. As such, the SIP covers the first part of the recommended course of action for which a reliable financing scheme was developed. The Short and Long Term Measures project findings served as a baseline for the decision to prepare and implement the SIP [1].

Implementation of the SIP began in 1998. A first milestone was reached in 2008 in terms of safety with completion of stabilization actions at the Shelter Object. In 2007 the contract for the NSC was awarded and work initiated. The main purpose of the NSC is to protect workers, the public and the environment from the impact of nuclear and radiation hazards associated with the Shelter Object and to provide conditions for further dismantling works and removal of FCM and radioactive waste.

Prior to emplacement of the NSC, several other activities had to be completed, with the aim of improving the overall safety conditions at the Shelter Object. These included creation and operation of the Integrated Automated Monitoring System, which includes several sub-systems designed to monitor radiation and nuclear safety, as well as seismic activity at the site and the structural stability of the original Shelter Object. In addition, several existing systems were modernized, including systems for physical protection, dust suppression and fire protection.

Specific to the final positioning of the NSC over the Shelter Object a large number of preparatory activities also needed to be completed, such as dismantling of 'old' structures and components (e.g. the old ventilation VT-2 stack), removal of dismantled materials from accessible areas within the Shelter Object and construction of the NSC itself, at a safe distance from the Shelter Object. The NSC was moved into place over the Shelter Object in 2016 followed by the start of commissioning activities in 2017. Work on installation and commissioning of the NSC was completed in 2019.

4.2.1.4. *Challenges and future perspectives*

A number of challenges and related uncertainties remain that require careful evaluation, research, investigation and sophisticated decision making, and can have a significant impact on the future planning and implementation for Phase 3, Conversion into an Environmentally Safe Site:

(a) Management of FCMs. FCMs need to be fully characterized as to their nuclear, physical and chemical states. Methods and technologies need to be developed to remove the FCMs from the site. These activities can be best supported by the development of a comprehensive R&D programme before on-site collection and removal is initiated. The location and configuration of FCMs inside the Shelter Object, the high exposure doses and the lack of access to certain portions of the structure, will create technical challenges for the removal of accumulated and identified FCMs.
(b) Management of 'fuel dust'. In addition to existing fuel dust resulting from the accident, the collection and removal of FCM will require disruptive and destructive techniques; these techniques will also generate 'fuel dust'. A specific area of concern is related to very small particulate dust, the potential for contamination spread and human uptake. The required studies and planning activities needed to address fuel dust issues will require comprehensive examination and study as part of the R&D programme. The development of proper implementation methods as well as measures to reduce worker exposure risks and ensure radiation protection are key considerations in future planning activities.
(c) Management of contaminated water and liquid radioactive waste from the shelter object. A facility for the separation of transuranic radionuclides and organic particles from water produced by the Shelter Object is not currently available but would be an essential step to make possible the treatment of this waste stream by the existing liquid radioactive waste treatment facilities.
(d) Dismantling of the existing Shelter Object. Unstable structures in the Shelter Object require dismantling. However, this activity also risks an increase in the level of airborne radioactive contamination and exposure doses inside the NSC. Moreover, dismantling of the radioactive contaminated structures and components will generate radioactive waste, including debris with FCM that will require removal and subsequent storage/disposal. To address the related needs, additional waste management infrastructure will need to be readied before significant steps to support dismantling of the existing Shelter Object.
(e) Additional facilities for radioactive waste management. As mentioned above, dismantling the Shelter Object and removal of FCM and other radioactive waste will create additional waste streams requiring processing, storage and ultimately disposal. For LLW, additional facilities will be constructed and operated near the Chornobyl NPP, while other facilities for higher activity wastes, such as deep geological repositories for ILW and/or HLW, remain to be developed. Due to the associated costs and planning requirements associated with geological disposal facilities, a dedicated disposal facility for the higher activity, longer lived accident waste from the Chornobyl NPP is not a realistic option. Disposal of these wastes can only be considered together with the national inventory of ILW and HLW/SNF. Therefore, planning activities related to the management of those accident wastes requiring geological disposal will require close coordination with the national waste management organization, which is responsible for developing geological disposal facilities, and integrated into the National Radioactive Waste Strategy of Ukraine.

The following general conclusions can be drawn regarding planning for the further implementation of measures to address the future safety of the Shelter Object:

(1) The Shelter Object under the NSC remains a damaged nuclear facility. The NSC is a means to provide safer conditions for the public, workers and the environment while decommissioning activities continue, and is not an end in itself. The situation at Chornobyl Unit 4, although safer, remains hazardous and considerable efforts, including investment of resources and time, will be

required before a truly environmentally safe condition can be achieved. While the NSC is equipped for the dismantling and removal of FCM and other radioactive waste, many challenges remain that need to be solved. These challenges and associated risks and uncertainties specific to FCM and radioactive waste removal require careful evaluation to allow appropriate planning, development and ultimately implementation of all necessary actions.

(2) A re-evaluation of the current long and short term strategies for the Shelter Object conversion is required in order to adequately reflect the scope of work anticipated over the 100 year lifetime of the NSC, including the development of a detailed road map of future activities. To this end, a staged approach that provides for future flexibility in design options can be considered. International cooperation and continued expert exchanges will remain critical in planning and subsequently implementing a recovery strategy consistent with Phase 3, Conversion into an Environmentally Safe Site.

4.2.2. Example 2: the Fukushima Daiichi NPP roadmap

The Mid- and Long-Term Roadmap towards the Decommissioning of TEPCO's Fukushima Daiichi NPP [59] depicts the phases of on-site activities. The roadmap is frequently updated and contains three phases, outlined briefly below:

(a) Phase 1:
- Period of stabilization and preparations for spent fuel retrieval from the Unit 4 spent fuel pool;
- This phase ended on 18 November 2013, with the start of spent fuel removal from the Unit 4 spent fuel pool.

(b) Phase 2:
- From the end of Phase 1 to the beginning of fuel debris retrieval;
- Target period: ~10 years;
- Including R&D activities, engineering planning and investigations of internal situations;
- Progressive management and treatment of stagnant water in buildings and spent fuel pools.

(c) Phase 3:
- From the end of Phase 2 to the completion of decommissioning;
- Target period: ~30–40 years;
- Phase 3-(1) is the period from the start of fuel debris retrieval to the end of 2031;
- Completion of fuel removal from spent fuel ponds for Units 1–6;
- Begin of a trial retrieval of spent fuel debris, which will be increased gradually.

The NRA directed TEPCO to submit an implementation plan for measures to be taken at Fukushima Daiichi NPP [79] in November 2012. In response to this direction from the NRA, TEPCO put together an implementation plan [80] and submitted it to the NRA in December 2012. The implementation plan was reviewed by the NRA and approved in August 2013. A request for change to the implementation plan was submitted in September 2013 and was approved in November 2013. The implementation plan describes the overall schedule for installing the specified facilities, their design, security and risk evaluation. It also describes the fuel debris removal plans and addresses decommissioning. For the final disposal off-site of the Fukushima Daiichi NPP, the MOE assumes a multistep approach moving from the start of the ISF through R&D and option evaluation stages to site selection and eventual disposal, following an eight step plan over a thirty year time frame [81].

4.3. ADDITIONAL REMARKS ON THE POST-ACCIDENT PLAN

An effective post-accident waste management plan will develop and maintain life cycle schedule and cost estimates. This will be necessary to ensure that funding needs are identified and funding sources secured. Option identification, analysis and decision making are key focal points of the waste management plan. A technology option assessment was a key element of the TMI-2 remediation effort. The technical history of the TMI-2 cleanup provides an excellent summary of the decision making process used to identify and select technologies [2].

It is important that predisposal planning for the waste disposal phase (see the next cSection) is coordinated with the overall life cycle management and cleanup plans for the site. To this end the waste management plan can be used to set priorities focused on hazard and risk reduction. The waste inventory assessment, including forecasts of future wastes, will in turn provide the basis for predisposal waste management and set the timeline for current and future waste management actions. As with any waste inventory, planning needs to account for wastes generated during future activities and secondary wastes resulting from planned waste recovery and treatment/processing activities. The need for and timing of future facilities or infrastructure modifications can be identified based on these waste forecasts.

Effective waste management planning will also include the identification of external interfaces regarding regulatory acceptance and licensing. Communications with stakeholders, including the public, are key components of the plan that can provide a forum for open information exchanges. Routine updates on the plan, especially those that demonstrate tangible progress, will enhance public confidence.

Further, the plan can include provisions for R&D to address known and emergent technology gaps. Capabilities from external organizations, both national and international, can be leveraged for R&D support. Moreover, these entities can also be utilized for programmatic and technological reviews over the course of the cleanup effort.

Maintenance of the waste management plan as a living document with periodic updates is needed to reflect new and changing information needs and inputs. Routine review and revision of the waste management plan is to be expected due to changing priorities, variable conditions and emerging needs. Feedback regarding both positive and negative outcomes is recommended in programmatic reviews to ensure continued advancement in the cleanup effort.

5. IMPLEMENTING THE WASTE MANAGEMENT PLAN

Key lessons learned:

- Waste management requirements and technology selection will depend on the details of the accident scenario, in terms of volumes and types of waste generated and their geographical distribution.
- The basic steps of waste management are the same as for all radioactive wastes. However, the required scale may be very much larger and the implementation timescales shorter. There may be a requirement for significant transboundary cooperation.
- The extent of geographical distribution of the waste needs to be considered when choosing centralized, distributed or mobile facilities. Often, a combination may be required.
- Emergency storage will almost certainly be required before waste treatment and longer term storage or disposal infrastructure can be put into place.
- Reuse or repurposing of existing facilities could shorten time schedules and increase the number of waste management options available.
- The urgency of the situation may dictate compressed time schedules for design, licensing and construction of new or temporary facilities.
- Construction and/or operation of facilities on a contaminated site will result in additional radiation protection requirements for workers and will increase the amount of secondary waste produced (e.g. protective clothing).
- The project life cycle needs to consider the decommissioning of any new facilities at the end of their life (i.e. the number, size and complexity of the required decommissioning needs to be considered).

— Implementation of an integrated management system can contribute significantly to quality assurance and configuration management during design, licensing, construction and operation, as well as quality control and record keeping.

The waste management plan developed in advance and immediately after any accident will involves several steps, as illustrated in Fig. 9. The first requirement will be an estimation of the location, amount, radioactivity and type of wastes that will be produced as the accident consequences are managed. This estimation needs to be carried out in a structured characterization process that includes forecasting of identified and projected wastes in terms of expected type, quantities and radionuclide content.

An initial step in managing the accident response could be 'emergency storage' of certain wastes at or near the site of the accident, allowing initial characterization, pretreatment, sorting and segregation of the wastes. Emergency storage, in the context of this publication, is buffer (temporary) storage for some period before further treatment or conditioning of the waste can start. The emergency storage period will typically last at least until the initial emergency period is over and may continue until subsequent management steps and facilities are available. This can range from months to several years.

Characterization may occur at multiple stages of the waste management process, from initial rough sorting to more detailed characterization for packaging for storage and disposal. For simplicity, it is shown as spanning the entire life cycle. Similarly, transportation of raw or packaged wastes might occur between several of the steps, as different processes and facilities become available. Transportation of large volumes of untreated waste may be necessary and is a significant issue in its own right (see Section 10).

Depending on activity levels and radionuclide content, there may be opportunities to divert or recycle some of the wastes at the pretreatment or treatment steps. Depending on Member State regulatory systems, it might be possible to identify some large volumes of low activity contaminated materials that can be disposed of directly in available landfills, or otherwise discharged as EW. For example, the US NRC has recently issued guidance on alternative disposal of VLLW in non-traditional facilities [82].

Appropriate treatment processes, packaging types and storage methods and technologies will need to be selected for the various waste streams and types. As well as being appropriate for the waste stream under consideration, the selected technologies need to be versatile, robust and cost effective, if possible, to handle multiple waste characteristics. Selection and deployment may be done incrementally (e.g. to reduce the waste volume, stabilize the waste and then package it for storage/disposal). In some cases,

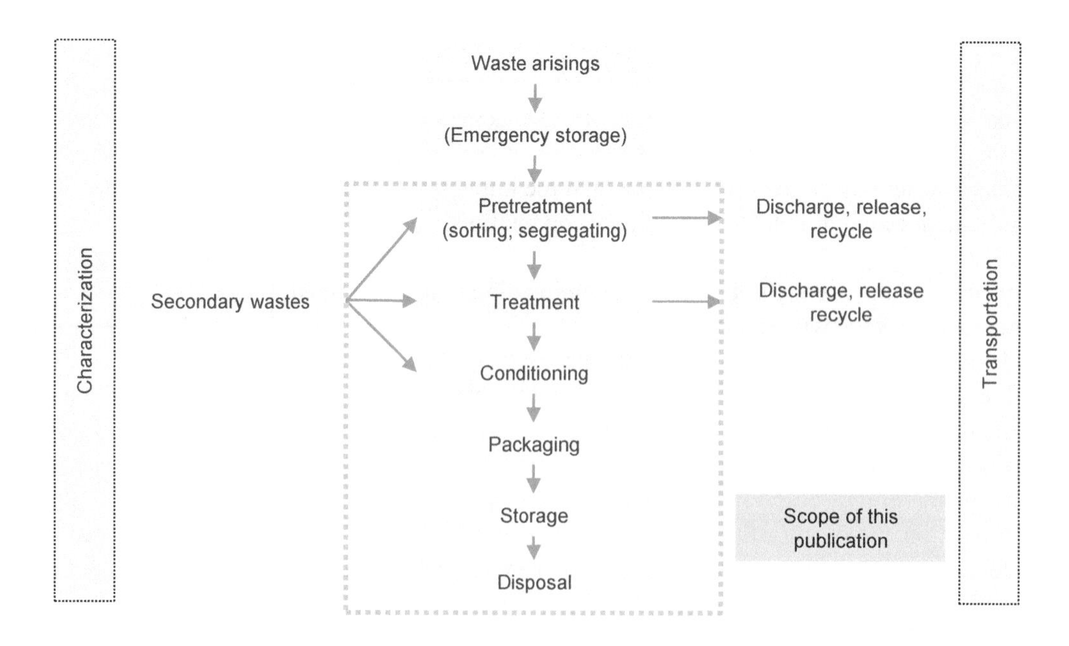

FIG. 9. General outline of the main steps in an accident waste management plan.

speed of deployment can also be a priority. In making decisions on options it is important to consider the entire waste life cycle, and to ensure that the selected options do not preclude future management steps.

The treatment step can itself create secondary wastes that will need to be managed and the balance of short term process optimization against longer term disposability and exposure reduction needs to be considered when selecting appropriate treatment technologies. In an accident situation, preference is often given to solutions that can be deployed quickly, perhaps using simple technologies or existing facilities that can be readily adapted for the new purpose. Since the full range of wastes and/or its characteristics may not be known in the initial stages, the versatility of the technology in handling a wide range of wastes is also important. The large volumes of some accident related wastes can also dictate the location of processing facilities near the waste source to minimize the need for transportation.

Although the disposal routes and methods may not be known with certainty at the time the wastes are treated, it is important to consider the eventual disposal schemes needed and available during the planning and execution of predisposal management steps (see Section 11). Care is needed in the selection and/or control of treatment and conditioning processes when the exact waste acceptance criteria for the disposal facility are not yet known to avoid an overly conservative approach. However, general WAC and limitations for most types of repositories are well understood and can be adapted to accident related wastes.

It is very important to understand the necessity of this early linkage between predisposal and future disposal solutions. This is especially true regarding optimization of disposal methodologies, since different options may be applicable, depending on the waste isolation and containment requirements that could be considered for essentially the same waste streams (i.e. deep geological disposal for HLW or ILW, disposal at intermediate depth or near surface disposal for VLLW or LLW). Therefore, predisposal options need to be optimized in consideration of the potential disposal solutions that can be pursued (see Section 11). Issues can also arise with the eventual retrieval of the stored wastes, especially if originally stored in raw (unprocessed) form.

Almost all of the steps shown in Fig. 9 will involve the use of infrastructure facilities for treatment, processing, storage or disposal of wastes. Some of these will already exist but could need adaptation — others might make use of repurposed infrastructure and it is almost inevitable that some entirely new facilities will need to be designed, sited and constructed. Implementation of all of the facilities will need to take licensing into consideration.

Based on the estimated amount and characteristics of the wastes, it is important to assess the possibility of using or adapting existing facilities. If the existing radioactive waste management facilities do not have adequate capacity or capability to manage the wastes and/or they have been damaged by the accident, new or alternative facilities will be required on either a temporary or a long term basis.

5.1. CONVERSION OF EXISTING FACILITIES

The conversion of existing nuclear and non-nuclear facilities from their original purposes, when possible, can prove beneficial in addressing waste management needs (Section 4.1.1). For example, non-radioactive municipal waste incinerators could potentially be used for the incineration of very low level organic debris material (such as wood, trees, foliage, etc.) resulting from the cleanup of contaminated areas. This might require certain modifications, such as adding monitoring systems on the exhaust or adding enclosure systems to contain radioactive ash, among others. Further, conventional and hazardous waste landfills can be equipped with leachate and gas generation control systems for the disposal of low level contaminated material and VLLW. Other examples of the use of existing facilities are provided throughout this publication. As with a newly built facility, applicable licensing regulations and procedures will need to be followed to obtain the necessary licences for conversion and radioactive operation.

While repurposing existing buildings as new waste management facilities may offer schedule and cost savings, it is also possible that legacy issues associated with those facilities may not be immediately evident.

The disadvantages of older and decommissioned nuclear facilities are natural deterioration of structures and degradation caused by abandonment of maintenance or in-progress decommissioning. Older facilities can be evaluated to determine whether they can be renovated in a cost effective manner to comply with current regulations, codes and operational standards. Operational and maintenance costs may also be higher than similar comparable modern buildings. These are factors that can work against the choice of redevelopment and reuse [83].

5.2. SITING NEW FACILITIES

If nuclear facility specific preplanning has been carried out (Section 4.1.2), then one of the outcomes will have been a list of options for locating various types of waste management infrastructure at or near the nuclear site. It is important to have a range of options available, as, depending on the nature of the accident, some of these locations may prove unusable, if infrastructure is damaged or heavily contaminated.

In addition to the conventional technical and non-technical siting factors for nuclear facilities, some specific considerations in siting new accident waste management facilities will include:

— Whether the facilities are temporary (e.g. stores) or permanent (disposal sites) and the timescales involved;
— Accessibility and ease of flow of materials from one waste management facility to another;
— Impact on future nuclear site management planning options;
— Acceptability of developing new facilities off-site against maintaining all waste management infrastructure on-site.

The specific considerations in siting new permanent disposal facilities are dealt with in Section 12.

5.3. DESIGN

Once a location and basic technology have been selected for a waste management facility, the detailed design for construction can be expedited. Compared to a design project without accident related urgency, the schedule for a facility to manage the early phases of an accident might be greatly compressed. This might mean, for example, a reduction in the number of design approval cycles. Grading the implementation of quality programme features can be considered, although this needs to be approached with caution to ensure that process barriers to poor quality are not adversely degraded and that the quality of the design, procurement or field construction and commissioning, especially for safety related systems, is not compromised to meet an aggressive schedule. Reliance on commercially produced items with reasonable levels of quality justification, such as fire protection equipment and electrical gear that complies with national standards, can minimize costs and acquisition times while providing adequate levels of performance.

In addition to the compressed schedule, there are also certain specific design requirements and features that need to be considered when dealing with accident related wastes. For example, the designs being evaluated need to:

(a) Address a wide range of wastes, often with characteristics that are not well defined. This suggests a preference for simple, robust and mature technologies. Use of technologies that are highly sensitive to waste properties or composition is not recommended.
(b) Consider uncertain endpoints for wastes, including secondary wastes (e.g. unknown disposal routes). Therefore, it needs to be flexible and provide sampling points throughout the process for determining product characteristics.

(c) Include adequate shielding of equipment and facility (including against possible effects of very high external radiation conditions) as well as consider the use of remote technologies and/or automation in order to reduce worker dose and increase throughput.
(d) Acknowledge interdependences and considerations of next steps (even if they are not fully known or well defined) such as provisions for sampling, emptying tanks and vessels, decontamination and decommissioning of the processing facility itself, etc.
(e) Include consideration of the eventual decommissioning of the facility, for example by providing access for decontamination, using materials that can be easily decontaminated, etc.

In addition, the design process needs to include clear understanding of applicable regulations, codes and standards, especially when involving multiple contractors and multinational teams.

5.4. LICENSING

The licensing process for an adapted or new facility will depend on the regulatory regime present in the affected Member State. There are several flexibilities that can influence decision making, both in preplanning for accident response and in the recovery phases. These include:

— Pre-licensing (or use of previously licensed) generic designs (e.g. processes, containers, storage concepts, etc.);
— Early determination of safety bases and environmental management needs and timelines for accident situations;
— Potential for modifying regulatory requirements in consideration of the accident situation, for example a streamlined process for accident phases; consideration of short term needs and facility lifespans where appropriate; efficient integration of regulatory decision making with the complex accident response schedule, such that radiological and nuclear safety matters requiring prompt attention are addressed in a timely manner;
— Applicability of international guidance, such as that provided for management of extraordinarily large volumes of radioactive waste following an accident, as discussed in Ref. [21].

In many Member States, a series of licences or authorizations may be required for construction and operation of a nuclear facility from nuclear, environmental and other regulatory bodies with jurisdiction over design and construction activities. From the nuclear regulatory perspective, these can likely include a construction licence or permit (after detailed design is accepted), a startup licence (after the initial phase of commissioning and cold tests), an operating licence (after hot test completion and configuration management has been ensured) and finally a decommissioning licence (when the facility is no longer required). These authorizations may include formal licences issued by a regulatory body, as well as permissions issued by other corporate or governmental authorities and/or agencies. In many cases, an environmental assessment is also required. However, in post-accident situations, especially in the initial phase of implementation of predisposal activities and facilities, where urgency of implementation is an issue, close coordination between regulatory bodies with the licensee is prudent to streamline the process as much as possible, including, where appropriate and necessary, exemptions from requirements on a short or longer term basis.

For example, at the Fukushima Daiichi NPP the relevant laws and regulations are based on normal plant operational conditions that have been difficult or impossible to apply to the post-accident situation. The Japanese NRA designated the reactors at Fukushima Daiichi NPP as a Specified Reactor Facility, that is, a facility where a nuclear accident has occurred, and special regulations were put in effect. The NRA implemented a suite of special regulations after determining the safety needs of the post-accident conditions and required the operator (TEPCO) to apply implementation plans relating to the facility operations. Off-site at the Fukushima Daiichi NPP, decontamination activities are carried out under the cognizance of the local government and ordinances for the regional contamination areas under the authority of the MOE [84], which has responsibility for the special contamination area.

A specific consideration for accident related wastes is the possibility of multiple Member States being affected by an accident. For example, several neighbouring states were greatly affected by the Chernobyl accident and its cleanup. This will inevitably lead to a number of national regulatory and executive organizations being involved, each of which may have substantially different standards and licensing requirements. Even within a given Member State, different requirements may be imposed at national, provincial/state and/or local levels.

5.5. FACILITY CONSTRUCTION AND PROJECT EXECUTION

Once a design has been selected, construction and preparations for operation can commence, with either prelicensing or concurrent licensing. Most Member States have the necessary infrastructure, expertise and personnel to carry out the nuclear construction projects that may be required. Such expertise is also openly available on the international market if a Member State needs additional assistance.

In the post-accident environment, the execution and management of construction activities will require consideration of the following:

(a) Establishment of a construction management team with clearly defined roles and responsibilities that is well integrated into the existing response team, including the existing operating staff of the nuclear site, to ensure sufficient coordination of construction in the post-accident environment.
(b) Access to staff covering multiple technical disciplines with adequate skills in order to ensure qualified support for the design/procure/construct/commission process in areas that involve specialized technical knowledge, such as waste treatment. Skills and abilities that are beyond normal nuclear facility knowledge will be essential.
(c) Proper and sufficient training in the conduct of work in a radiological environment, including a full understanding of all related logistical support needs.
(d) Measures to ensure the safe and efficient integration of a large number of short time frame projects and contractors, as well as interfaces with multiple other related response projects and responsible agencies.
(e) Accommodation to integrate international advisors, contractors and personnel (e.g. to manage language/cultural issues, including the need for interpreters and translation of work instructions and drawings).
(f) Diligent implementation of a nuclear safety culture and compliance with post-accident operational processes.
(g) Appropriate caution may need to be employed when following existing operating and engineering procedures that were valid for normal operation but may not be fully suitable for a post-accident situation.

Other factors that may be critical to the construction of the required facilities in a post-accident situation include:

— Availability of funding, construction materials, equipment, infrastructure and experienced personnel;
— Possible off-site prefabrication of major project subassemblies or components to minimize on-site construction time and/or radiation exposure.

Configuration management is an essential component not to be overlooked during the design, construction, startup and operation of the facility. Configuration management ensures that the 'as-built' facility matches the design documentation and licensing basis [85]. In past responses, this aspect has often been inadequately addressed or fully forgotten in the design and construction of facilities under aggressive response schedules, or where inadequate resources or funding were available. Maintaining both the as-constructed design configuration and the operating configurations consistent with the safety analysis is essential to assuring that the integrity of the radiological safety, nuclear safety, waste confinement and isolation features is maintained. For example, any physical modifications to the facility are subject to pre-implementation safety reviews and the facility has to be operated within the envelopes defined by operating, maintenance and test procedures that have been developed, consistent with the safety analysis.

In the compressed time schedules that might be associated with accident related facilities, construction might sometimes need to begin before all aspects of the facility are fully designed. In this case, it is very important that a highly disciplined and effective configuration management system is in place to ensure that the facility is constructed as per the design and that all changes have been properly documented. Ideally, the facility operator needs to be provided with complete and accurate documentation (drawings, manuals, etc.) of the facility design and construction prior to starting operation. Rigorous procedures for updates to as-built drawings and analyses are likewise essential to ongoing safe operation, particularly where they will introduce changes to operating, maintenance and test procedures.

5.6. COMMISSIONING AND TESTING

A newly constructed or adapted facility will almost always be subject to a startup commissioning and testing programme to ensure that all equipment functions correctly, and all essential operating and maintenance procedures are consistent with the design and safety analysis. A commissioning plan specifies the tests and performance criteria that are to be met to demonstrate that the facility has been constructed and can operate in compliance with its design bases. Commissioning normally has a cold test stage (non-radioactive, simulant based) and a hot (radioactive) test stage, with the facility then transitioning into a full capability operating phase. This transitioning process is important in validating the efficacy of operating, maintenance and test procedures, and as a training vehicle for operational phase personnel working with new equipment. In the post-accident scenario, it could be expected that the hot stage will be used to turn over the new or reconstructed facility to its operating crew.

In a post-accident situation, the time available for commissioning tests may be limited and the performance criteria may not be fully defined (e.g. due to the broad characteristics of the waste). Additionally, issues can arise related to scaleup from bench scale tests, or oversights that might be due to the compressed initial design schedule. These need to be corrected prior to radioactive operation. Therefore, it is important to carry out these tests to the extent practical, consistent with regulatory authority requirements, to ensure that the facility is operating as per design before large scale radioactive operation is started.

5.7. TRAINING

New or adapted waste management facilities and their associated technologies are likely to require specific training programmes both for existing nuclear site staff and new staff brought in as part of the accident response programme. Nominally, staff training begins no later than during the commissioning phase of a facility, well before it is ready for operation. It is important that the training fosters implementation of a strong nuclear safety culture that includes rigorous compliance with facility procedures and a predisposition toward safety over production.

Involvement in the commissioning programme or secondment to similar facilities can be used to supplement classroom style training. The commissioning phase provides a useful opportunity for familiarization with the facility and its procedures. The availability of suitable training facilities needs to be considered in the planning. If none is available, they will need to be constructed as part of the facility project and made available prior to the completion of the main facility.

5.8. OPERATION AND MAINTENANCE

Many technologies and types of facility that will be employed for post-accident waste management are well understood and have been widely used in other non-accident applications. However, operating and maintaining a waste management facility in a post-accident situation may pose some challenges. Familiar operating, maintenance and test procedures and management methods could require adaptation, by either modification of existing facility procedures or creation of new procedures to accommodate the post-accident conditions. This will require consideration of various factors, including:

- Some post-accident situations require waste processing on a much larger scale than has been employed in the past, leading to much larger equipment or more numerous modular facilities;
- A need for flexibility in operation to accommodate varying waste feeds to processes (e.g. it may require more manual intervention to control process parameters than is normal);
- Management of working times (e.g. due to environmental conditions and/or higher radiation dose rates) and associated prioritization and sequencing of activities;
- Adaptation of staffing levels and shift rotations to accommodate work times and operational throughput requirements;
- Adaptation of maintenance practices to accommodate difficult conditions (e.g. radiation fields, airborne contamination, absence of complete system or component documentation, urgency for facility return to service, replacement versus in situ repair, etc.);
- Conditions may warrant use of remote operation and specialized technology in order to manage worker dose exposure, which will also require specialized procedures, training and management processes prior to implementation;
- Operational procedures for a new or adapted facility could have associated needs for logistical support, including supply chain support for material requirements, on-site logistics and staff support such as worker amenities (e.g. changing rooms, decontamination centres, living quarters, transportation, support facilities, etc.).

5.9. DECOMMISSIONING OF WASTE HANDLING AND STORAGE FACILITIES

Any facilities used for the predisposal management of radioactive wastes will eventually reach the end of their service life, either by completing their purpose or through retirement due to age or obsolescence. Decommissioning will generate additional wastes and worker doses. Therefore, careful consideration of the number of facilities being utilized is needed. Further, the minimization of secondary wastes and ease of decontamination are important factors for inclusion in the designs of any new facilities.

It is noted that the post-accident situation could last much longer than the typical design life of waste management facilities. It is important to take this aspect into consideration in the early stages of decommissioning planning so that either capital reconstruction or replacement of equipment, or even completely new facilities, can be envisaged in time. Early decommissioning of waste facilities may also be triggered by the application of more advanced and efficient technological solutions, or by better understanding of the characteristic of waste streams that need to be tackled over time. For example, retrieval and management of fuel debris from the Chornobyl Shelter may occur in a 50 to 100 year time frame and will require the design and construction of specialized waste management facilities, while at the same time all existing processing facilities installed at Chornobyl NPP will be well past their design life.

Decommissioning of nuclear facilities is a well understood practice in most Member States with nuclear power or advanced nuclear research capabilities. In the case of facilities used for predisposal management of accident related wastes, especially those that were not originally designed as nuclear facilities, special care will be required to control the spread of contamination during equipment and building dismantling.

As with any nuclear decommissioning project, a decision will be required on the end state of the facility (i.e. defined future use conditions). In the case of a post-accident situation, the end state criteria need to be coordinated with the end state for other accident related facilities and areas, and a consistent approach needs to be taken. For example, it is not realistic or practical to devote resources to achieving a much higher level of decontamination of a decommissioned waste management facility than that of the surrounding site, if that will remain contaminated from the accident itself.

5.10. QUALITY ASSURANCE AND QUALITY CONTROL

Quality assurance (QA) is part of an integrated management system that includes all those planned or systematic actions necessary to provide adequate confidence that a product or service will be of the type and quality needed and expected by the customer and other stakeholders. In conjunction with QA, quality control (QC) is a quality confirmation or confidence building function that includes all those planned or systematic actions (e.g. scientific precautions, such as calibrations, duplicate samples, sampling methodology, chain of custody, physical inspections, non-destructive examination, etc.) that are needed to provide confidence that a product or outcome actually meets specifications. QA complemented by QC is an important element of a waste management system. Basic quality requirements are specified in other publications, such as Ref. [46].

Stakeholders, regulators and other governmental bodies rely on quality programme documentation to provide confidence in any nuclear project; the quality programme itself places considerable requirements on the documentation to act as evidence of satisfactory products and activities. Consequently, it is important to create and preserve the documentation of quality achievement.

In a post-accident situation, it is expected that some documentation that is required for the design or operation of a waste management facility and dates to the earliest phases of the accident event, especially emergency phase records and early waste related activities, may be missing or otherwise incomplete. Important records include those on types and quantities of waste, their characteristics, how they were treated and/or packaged and where they are stored or were disposed of. Under such circumstances, it may be necessary for a facility management team to recreate or replace missing information to ensure that appropriate levels of quality and safety are re-established. This could involve the reconstruction of documentation, where possible, or the reperformance of activities to re-establish records as soon as is practical.

6. WASTE CHARACTERIZATION STRATEGIES, METHODOLOGIES AND TECHNIQUES

Key lessons learned:

- Characterization of wastes affecting facilities and off-site locations is required for waste management (including for clearance), remediation, cleanup and personnel protection purposes, as well as for decision making.
- Prompt acquisition of characterization data underpins technical decision making, planning and engineering. This is an iterative, ongoing process.
- To optimize sampling, characterization strategy, goals and objectives need to be clearly understood and documented before embarking on a characterization programme.
- Additional laboratory facilities and/or analytical techniques might be required as a result of an accident (e.g. located in an area of low background dose rate).

— Situation specific radionuclide vectors or scaling factors might need to be developed for a variety of non-standard waste types.
— Consideration has to be given to data management in terms of compiling, organizing and disseminating technical data.

6.1. CHARACTERIZATION STRATEGY AND METHODOLOGY

IAEA Safety Requirements stipulate that characterization applies to each step of radioactive waste management. Characterization is also applicable to waste generated by a nuclear accident [17–19]. A graded approach can be adopted, meaning that the degree and effort of characterization necessary for the purpose is a consideration. When a nuclear accident occurs, the challenge is to be able to mobilize appropriate expertise and equipment over a short timescale to measure the spread of radioactivity in a variety of pathways and media, recognizing that protection of the population is the prime objective. Therefore, during the initial phase of a nuclear accident, waste characterization will focus on assessing dose levels to determine the magnitude of the accident. These data will facilitate the temporary staging and sorting of waste as well as informing event management about required protective actions. When plans are being developed for subsequent waste treatment, it will be necessary to provide more detailed characterization data on radionuclide type and concentration, chemical composition and physical properties, to inform processing, storage, conditioning/packaging and eventual disposal decisions.

6.1.1. Process knowledge

In the event of a nuclear accident, knowledge of the accident scenario, the design of the facilities involved in the accident, the available waste management capabilities, the prevailing environmental conditions, the materials and the radionuclide concentrations released, all contributes to the determination of the sample acquisition and measurement methodologies needed. This is particularly important for radionuclides that are difficult to sample or measure. Characterization needs after a nuclear accident vary widely, ranging from the need to assess the radiological impact over a large geographical area to the need for quantitative radionuclide data to support performance assessment calculations to facilitate waste disposal.

6.1.2. Establishing a characterization programme

The post-accident cleanup phase commences once (or even before) the accident has stabilized. A characterization programme is required to categorize the waste and ensure that it meets both the safety and downstream waste acceptance criteria requirements. Important factors for consideration by the characterization programme are:

(a) Goals, objectives and the characterization strategy need to be clearly understood and documented before embarking on a characterization programme, so as to avoid unnecessary sampling and its associated worker dose.
(b) Characterization priorities may need to be set to address some especially problematic parts of the accident site early, before taking steps for dose reduction or release mitigation; for example, removal of hot spots or consideration of radiation area stay times or other environmental working condition restrictions.
(c) Requirements for logistical support, including sourcing appropriately skilled and trained staff, instrumentation and (mobile) laboratory support, material support (sampling consumables, personal protective equipment, etc.) and staff support amenities (e.g. change-rooms, decontamination centres, living quarters, transportation, support facilities, etc.).

(d) Potential for using remote controlled unmanned aerial vehicle (UAV) technology to reduce worker exposure to gain access in areas with harsh radiological or other hazardous conditions; for example, aerial drone survey, small and large sized remotely operated vehicles and robots.
(e) Requirements to shield characterization instruments, equipment and facilities for management of background radiation interference and for worker protection.
(f) Maintenance capabilities for characterization equipment and facilities under difficult conditions (e.g. decisions on whether to replace versus repair in situ).

When identifying appropriate sampling and characterization techniques and technologies it is important to consider the following:

(1) The strategy for sampling waste materials:
   - The range of materials that need to be characterized, in terms of their physical, radiological and nuclear properties (raw waste, large objects, bulk materials, debris, vegetation, soils, special nuclear material, etc.);
   - The extent, frequency, accuracy and precision needed in sampling, including the selection of sampling points and any limitations in accessing materials that require characterization;
   - The sampling limitations resulting from the sampling environment (worker dose, ambient environmental conditions, physical location).
(2) The measurement limitations resulting from the sampling environment (radiation levels affecting the minimum detectable activity, etc.).
(3) The use of specific and graded approaches for dealing with uncertainties and the effects on sampling representativeness; for example, approaches used in normal operational modes or decommissioning might not be applicable due to the special characterization needs for accident waste storage, treatment and disposal.
(4) The need for special measurement techniques; for example, for alpha and pure beta characterization in accident situations (e.g. required sample size, background gamma levels, detection limits, wet chemistry, specialized facilities required, such as off-site transport of samples, etc.), plus any specific requirements for calibration of instruments.
(5) The use of indirect characterization methods, such as non-destructive assay (NDA) versus destructive assay (DA) and differences in on-site versus off-site sampling needs, as sampling plans will not be defined on the same basis for NDA (e.g. in situ gamma spectrometry and DA).
(6) The use of scaling factors; where appropriate, considering the potential volume and heterogeneity of waste streams, it may be necessary to scale up from limited sampling and characterization data, based on process knowledge for the accident facility.
(7) The use of data quality objectives (DQOs; see below) and the application of QA and QC processes to sampling and analysis, including to the less direct measurement technologies, such as NDA and DA, including the application of effective processes to ensure data and records integrity and provide for:
   - Traceability and accuracy for records and data capture;
   - Physical (and electronic) protection and appropriate cataloguing of records and data in storage, including provisions to ensure that the latest information is available and provided to users;
   - Control over the dissemination of records and data, to ensure their delivery to users and accountability for their use.

The use of DQOs and the application of plant process knowledge are described below.

6.1.3. Data quality objectives

DQOs are qualitative and quantitative statements that define the type, quality and quantity of data necessary to support defensible risk management decision making. The DQO concept was developed by the US Environmental Protection Agency and was designed for use in developing an effective sampling or testing plan that optimizes the data that are to be collected [86]. However, the principles of DQO can be applied to other aspects that require decision making. For example, Sellafield Ltd (UK) has applied DQO to define R&D needs.

DQOs are developed prior to collection of data, as part of a sampling/characterization programme, in order to provide a logical structure that focuses data collection on information needs. Specifically, DQOs ought to:

(a) Clarify the data collection objectives;
(b) Specify the use of the data and how the data will support the risk management decisions;
(c) Define the types of data to be collected;
(d) Specify the quantity and quality of data to be collected.

Understanding what information is already available and can be used may reduce sampling/characterization effort. Use of the DQO process leads to efficient and effective expenditure of resources, consensus on the type, quality and quantity of data needed to meet the goal, and the full documentation of actions taken. The DQO process is iterative. As assumptions or data needs change, or where findings influence new data needs, DQOs will require reconfirmation and, if necessary, revision, as illustrated in Fig. 10.

The DQO process consists of the following seven steps:

(1) Step 1. State the problem. Clearly define the problem that requires new data so that the focus of the study will be clear and unambiguous.
(2) Step 2. Identify the decisions/goals of the study. Define the decision that will be resolved using data to address the problem.
(3) Step 3. Identify information inputs. Required to resolve the problem, and to determine which inputs require data measurements.
(4) Step 4. Define the boundaries of the study. Specify the spatial and temporal circumstances that are covered by the problem.
(5) Step 5. Develop an analytical approach. Integrate the outputs from previous steps into a single statement that describes the logical basis for choosing among alternative actions.

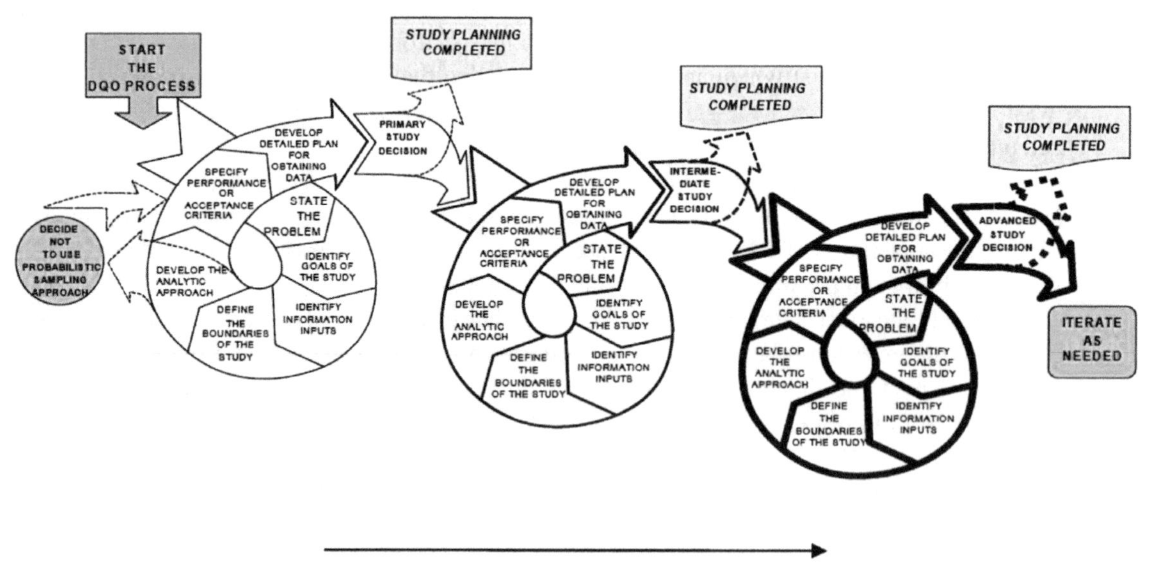

FIG. 10. Document quality objectives process iterations through a project life cycle. Courtesy of the US Environmental Protection Agency [86].

(6) Step 6. Specify the accuracy and precision required and include these in performance or acceptance criteria to support decision makers' understanding and consideration of data uncertainties.
(7) Step 7. Develop plan for obtaining data. Identify the most resource effective sampling and analysis design for generating data that are expected to satisfy the DQOs.

6.2. VARYING DEMANDS FOR CHARACTERIZATION DATA

Many types of materials will require characterization, at a wide range of spatial and volume scales, for a range of end users and purposes. This section looks at some of the varying demands for characterization information that are likely to arise after an accident.

6.2.1. Bulk characterization

A severe nuclear accident can result in contamination over a vast area. Contamination levels will need to be assessed at an appropriate scale to ensure public and environmental safety and to support developing cleanup strategies. Coordination efforts with neighbouring countries may also be required.

The combination of radiation detection and global positioning satellite (GPS) technology has greatly enhanced our ability to characterize radiological conditions over geographically large areas. Crewed aerial vehicles, land vehicles and on-foot radiation detection measurements have all been deployed following NPP accidents. Advances have also included the development of unmanned aerial vehicles (UAVs; drones) and remotely operated surface vehicular systems.

For example, a field monitoring vehicle was developed for the Fukushima Daiichi NPP, for use in dose rate and contamination surveys of ground areas, including roads, playgrounds, school yards, sports areas, parks, etc. [87]. The system is based on a Japanese agricultural tractor with a trailer. The system includes hydraulic dampers to accommodate irregular surfaces, for example bumps and vibrations. When detected above designated threshold values, a built-in painting device automatically marks the location of contamination. The detection technology used is based on Geiger–Müller detectors for dose rate measurement, a GPS for localization and a 1000 $mm^2 \times 500\ mm^2$ plastic scintillator for surface contamination measurement. The detection unit is protected with lead shielding to permit detection of low level surface contamination, even in areas with a high background. The detection threshold is approximately 1000 Bq/m^2 to 2000 Bq/m^2 in a natural background of ~0.1 μSv/h.

6.2.2. Characterization for sorting and staging

Sorting wastes is a prudent measure to support effective temporary storage, manage dose rates, stage for subsequent processing, package for disposal, etc. In the early stages of accident recovery, dose rate measurements can be used to segregate wastes for temporary storage and enhance worker safety by storing waste materials with higher dose rates away from work activities. Once treatment processes are identified for specific wastes and characterization data are obtained, the wastes can be staged appropriately into treatment queues. Characterization data on specific attributes of the wastes are needed to ensure appropriate wastes are treated.

6.2.3. Characterization of bags and packages

Temporary storage of accident generated waste needs to be judiciously planned and implemented, often very quickly following the accident, to reduce public and worker radiation exposures. The type of container used depends on the radiation levels, chemical nature and physical properties of the waste. Containers can include bags, drums, boxes, HICs, etc., of various shapes and sizes. Characterization information is needed for proper selection of temporary storage packages to support storage area/facility dose rate and contamination control and for inventorying and subsequent handling and disposition. In the

Fukushima Daiichi NPP off-site case, the 1 cm surface dose rate is measured for every flexible container and the 1 m air dose rate is measured after loading them on to trucks, as shown in Fig. 11.

6.2.4. Release, clearance and recycling of materials

Processes have been developed to exempt materials, based on radiation clearance levels. Clearance processes initially begin with specification of materials under consideration for release from regulatory control and end with any follow-on actions resulting from monitoring. In order to ensure compliance with exemption and clearance levels it is necessary to establish a monitoring programme. The process for this compliance monitoring, when compared to radiation monitoring at operational facilities, can be very complex, involving time consuming procedures. Monitoring clearance levels typically involves measuring large quantities of materials. However, the clearance levels are near the detection limits of field equipment and confirmation sampling at suitably equipped laboratories may be required [35].

Fundamental measures have been instituted at the Fukushima Daiichi NPP to address the increase in accumulated water due to groundwater inflow and to prevent inflow into the reactor buildings [80]. The operation of 12 groundwater bypass pumping wells commenced sequentially to intercept and pump up groundwater before it could reach the damaged reactor buildings. The pumped up groundwater was first temporarily stored in tanks with releases to the sea beginning in 2014, after TEPCO and a third party organization confirmed that its quality operational targets were met [89].

	Mean radioactivity concentration (Bq/kg)						Maximum equivalent rate 1m from the truck (vehicle transfer rules)
	3,000	8,000	30,000	150,000	500,000	1,000,000	
Air dose rate (μSv/h)	0.27	0.72	2.7	13	44	89	100

FIG. 11. Example of radiation measurement for a bag (top) and a truck at Fukushima (bottom). Adapted from Ref. [88].

At the Chornobyl NPP, free release of slightly contaminated accident material was not pursued, due to the establishment of the exclusion zone surrounding the accident site. Materials from decontamination efforts related to the accident were buried and marked with signage. The total area of the exclusion zone has full access control. Inside the boundaries of the exclusion zone , a further restricted area was established to encompass all temporary waste storage locations where contaminated material was buried.

6.2.5. Food supply characterization

Three main pathways exist for post-accident radioactive substances to enter the food supply: they can be carried by the wind (i.e. airborne) or be carried by water or through soil and sediments. Radionuclides in air can be deposited on plants that are eaten by animals. Radionuclides in water bodies can also affect aquatic plants and animals in a similar way. Contaminated food supplies and other biological material might need to be treated as radioactive waste, depending on the level of contamination. Therefore, characterization of food supply could be required for waste management purposes, as well as for human health protection.

Radionuclide activity concentrations in the environment serve as an important initial criterion for identifying future agriculture and food production management options. Early on in the accident response and as an initial first step it is important to assess how contamination has spread in the environment by measuring and then mapping radionuclide concentrations. Radionuclide concentrations can be measured either in situ or in a laboratory.

Fukushima Prefecture is a very rich agricultural area, supplying high quality rice, beef and fruits to the rest of Japan. Before the accident, the limit for radioactivity in food was 600 Bq/kg total $^{134+137}$Cs across Japan. In March 2012 the limit was reduced even further to 100 Bq/kg. Purchases from Fukushima Prefecture declined, despite an extensive sampling and assay programme, which ensured that the release and consumption criteria were met. In order to further foster public confidence, every bag of rice from Fukushima Prefecture was assayed for radioactivity, not just a statistical sample. This same process was deployed for Anpogaki dried persimmons, a delicacy from the area. The drying process for Anpogaki persimmons increases the concentration of radioactivity in the fruit and measurements ensure that no persimmons in excess of the dose limits can enter the market.

To support food supply monitoring, several development projects started after the Fukushima Daiichi accident for verification of contamination level and to check whether food was acceptable for human consumption. The Canberra FoodScreen Radiological Food Analysis System is a laboratory grade gamma spectrometry food analyser based on a GeHP spectrometer for measuring contamination of raw or processed food [90]. The system is preconfigured and preloaded with generic efficiency calibrations covering a wide variety of raw and processed food sample containers and matrices. Specialized geometries are also possible.

After the Chernobyl accident, measurement systems were developed to screen for radioactivity (gamma emitters, such as caesium isotopes) in live animals before the animal entered the food chain. Monitoring of live animals can be carried out both at the farm and/or at slaughterhouses, providing high confidence that meat is safe before it enters the food chain. This approach has thus far not been successful in Japan, as the beef husbandry practices are different. However, a system was designed to measure radioactivity in the local hay and silage grown to feed the cattle. These are examples of customizing and optimizing methods for food control to ensure that samples can be analysed quickly and efficiently.

After the Windscale Piles accident, as a result of strontium releases, a large scale environmental monitoring programme was conducted and the results of this survey led to a restriction on the distribution of milk from an area adjacent to Windscale Works for a period of several weeks [91].

6.2.6. Underwater characterization

The radiological assessment of water bodies is also necessary after a nuclear accident to ascertain the safety of drinking water supplies for plants and animals living in affected water bodies, and for water use (e.g. for agricultural purposes). Heavily contaminated water bodies could also result in the generation of additional radioactive waste, either directly or via cleanup programmes.

Historically, in situ γ spectrometers have been designed for underwater operations [92]. The spectrometers initially contained NaI(Tl) detectors. Later, spectrometers using HPGe detectors with adapted electronics, data acquisition and processing electronics were also developed. Both systems are in wide use. Units are also available with a hydraulic winching system with hundreds of metres of conducting cable. These systems have been used for the measurement of radioactivity in seawater and monitoring of the seabed. Underwater γ spectrometry has been used for in situ monitoring of leakages of radionuclides from dumped or sunken nuclear objects/waste (e.g. nuclear submarines) or discharges from nuclear plants.

Remotely operated vehicles (ROVs), such as the one developed by AREVA [93] and equipped with gamma spectrometry, have been deployed for underwater inspection of lake beds in the Fukushima region to measure sediment contamination. The ROV is capable of operating down to a depth of 200 m. The detection technology is based on a $LaBr_3$ probe with countermeasures to handle water column shielding issues. The ROV is localized using an ultra short base line, which is an underwater positioning system that uses a vessel mounted transceiver to detect the range and bearing to a target using acoustic signals, linked to the boat at the surface. The position of the boat is determined using GPS.

6.2.7. Characterization for processing, storage and disposal

Waste generated from a nuclear accident will have varying physical, chemical and radiological (and perhaps nuclear) characteristics, posing special challenges for conditioning, packaging, storage, transport and disposal. Close coordination is essential between all those involved in the entire sequence of waste management, from designing the processes needed for waste conditioning through to establishing waste acceptance criteria for predisposal stages of waste management. The waste characterization programme has to recognize that the initial characteristics of the waste (as far as they are defined in earlier stages) could have changed over time as a result of various degradation processes, including biodegradation, chemical reactions, corrosion and radioactive decay. As a result, the original documentation associated with the waste package may not be sufficient, without additional information, to describe the current status of the waste. Moreover, there could be scepticism regarding the available data sources because different information pathways can lead to contradictory data.

As previously discussed, a suite of characterization data will be required to support subsequent waste treatment and disposal. The complexity and volume of the waste can drive the waste management programme toward a very large sampling and analysis effort. Therefore, it is essential that a sampling and analysis programme is established that provides the required data using the most efficient means possible. The DQO process discussed in Section 6.1.3 can be utilized to facilitate sample and analysis planning.

Both non-destructive and destructive characterization techniques will be required to support waste processing, storage and disposal. The measurements to be made and/or techniques to be used could include:

(a) Non-destructive assay:
    - Dose rate cartography and dose rate conversion methods;
    - Surface contamination with portable instruments;
    - Gamma spectrometry;
    - Neutron measurement (for dismantling of reprocessing plant, materials contaminated with Pu, Cm, etc.);
    - X radiography (to assess packages/containers containing radioactive waste).

(b) Destructive assay:
    - Radiochemical analyses;
    - Chemical analyses;
    - Rheological properties.

At the site of the accident, there will likely be only limited existing analytical capabilities and off-site laboratories will probably be required to support the analytical programme. However, depending on the size of the analytical programme, sample transport can become problematic and may dictate the need to establish specific analytical capabilities at or near the site. Radiation levels as a result of the accident can complicate the siting of a laboratory facility, due to the need for a low background dose rate area.

6.3. CHARACTERIZATION METHODS AND TECHNIQUES

Innovative characterization techniques and analytical methods have proven vital in the cleanup of past nuclear accidents. The deployment of mobile systems, the development of equipment to perform analyses in difficult environments and the development of an analytical method for characterizing complex matrices have all contributed to past waste management and decommissioning efforts. Today, these techniques and equipment provide a toolbox for future use.

This section provides an overview of the methods, techniques and equipment that have been deployed to support the characterization efforts after nuclear accidents.

6.3.1. Dose rate measurements

Dose rate measurements, in conjunction with airborne contamination survey systems, will provide a speedy indication of the severity of a nuclear accident. Such measurements can also be used to characterize materials, soils, buildings, sites, etc. after an accident.

Dose rate measurement can be performed via different types of detectors. Gas detectors such as Geiger–Müller tubes, ionization chambers or scintillators, either plastic or sodium iodide (e.g. NaI(TI)), can be used, depending on the dose rate level and measurement conditions. Dose rate monitors can be handheld or mounted on a vehicle (car, truck, helicopter, airplane, underwater remotely operated vehicle, robot, etc.) and can be used in conjunction with GPS systems to create dose rate or contamination level maps.

6.3.2. Contamination survey with alpha, beta and gamma probes

Alpha, beta and gamma probes will provide an indication of the level and type of contamination of soil, ground covers (e.g. streets, walkways, etc.), walls and other surfaces. A drawback is that monitoring takes time and workers perform the survey in hazardous radiological conditions. This type of equipment is only appropriate for use in low external dose rate backgrounds. Surface contamination of a container with waste inside cannot be assessed with such equipment. The smear test methodology is appropriate for such a purpose, with the smear being counted in a low dose rate area.

Detection systems are based on a scintillation counter with a large area scintillator window and integrated photomultiplier tube. A scintillator such as zinc sulphide can be used for alpha particle detection and plastic scintillators can be used for beta detection. The resultant scintillation energies can be discriminated so that alpha and beta counts can be measured separately with the same detector. Other measurement systems are based on a gas proportional detector. They comprise either handheld monitors or fixed monitoring equipment for area or personal surveys, or installed for personnel monitoring, and require a large detection area to ensure efficient and rapid coverage of monitored surfaces. These probes can be mounted on a remote controlled vehicle or robot.

6.3.3. Measurement of airborne contamination

In this context, airborne contamination refers to gaseous or particulate releases to the atmosphere. Airborne contamination survey systems are widely used for health physics purposes throughout the nuclear industry and can provide a rapid indication of the severity of ongoing accident events or recently stabilized accidents. In most particulate monitoring systems, air is drawn through the instrument by means of an external pump or vacuum system, and particulate material is deposited on a removable card mounted filter. The filter is monitored by a detector, for example a passivated ion implanted planar silicon detector, which allows simultaneous measurement of both alpha and beta radioactivity of the material deposited on the filter. The airflow is measured directly and reported by the instrument. The design of the airflow system is optimized to ensure high air sampling efficiency and high transmission of particles to the filter. The systems are usually used inside nuclear facilities but can also be used outside and they can be installed in mobile laboratories. Radioactive gas detection systems use similar sample capture airflow mechanics but read the gas activity levels directly.

6.3.4. Gamma spectrometry

Gamma spectrometry measurements can be undertaken on samples, waste packages and in the field, where they can provide a speedy indication of the severity and nature of a nuclear accident. Gamma spectrometry is a non-destructive method that is able to determine the intensity and energies of gamma rays emitted by the item measured. The collected and analysed gamma emissions provide a gamma spectrum that characterizes the radiological fingerprint of the measured items in terms of gamma emitters and makes it possible to evaluate the item's activity, with an adequate calibration.

Two types of detectors are used for gamma spectrometry, scintillators such as sodium iodide crystals (NaI) or lanthanum bromide detectors ($LaBr_3$) and semiconductors such as CZT (cadmium telluride implanted zinc) or germanium detectors. The resolution varies with the type of detector. A germanium detection unit includes a Dewar flask for liquid nitrogen or an electrical cooler. Gamma spectrometers can be handheld or mounted on a vehicle (car, truck, helicopter, airplane, underwater remotely operated vehicle, robot, etc.) and can be used in conjunction with GPS.

6.3.5. Gamma imaging

A gamma camera can be used after a nuclear accident to detect hot spots/activity concentrations, for improvement of modelling (when coupling with gamma spectrometry) and for planning of interventions. Significant experience feedback exists related to the use of gamma cameras in the management of historical legacy waste.

Gamma imaging consists of a visual image of the target area with a superimposed gamma profile showing the location of the radiation, with a graded colour scheme that indicates the magnitude of the radiation, as shown in Fig. 12.

6.3.6. Neutron counting

A neutron counting system measures the neutrons emitted from fissile material and other actinides and is used for nuclear material accountancy purposes. Neutron passive counting relies on an actinide fingerprint that is based on radiochemical analysis of samples. The source of neutrons cannot be determined simply by detecting them. Some minimal amount of knowledge of the waste stream characteristics is required. Most methods rely on ^{3}He proportional detector tubes or BF_3 filled detectors. Neutron counting methods are inherently complex characterization methods. There are no simple neutron methods — each requires a knowledgeable physicist to manage the measurement campaign to ensure accurate interpretation of results. Passive neutron counting systems are currently available for the characterization of volumes of less than 1 m^3 or 1.5 m^3, while R&D is ongoing to characterize larger volumes.

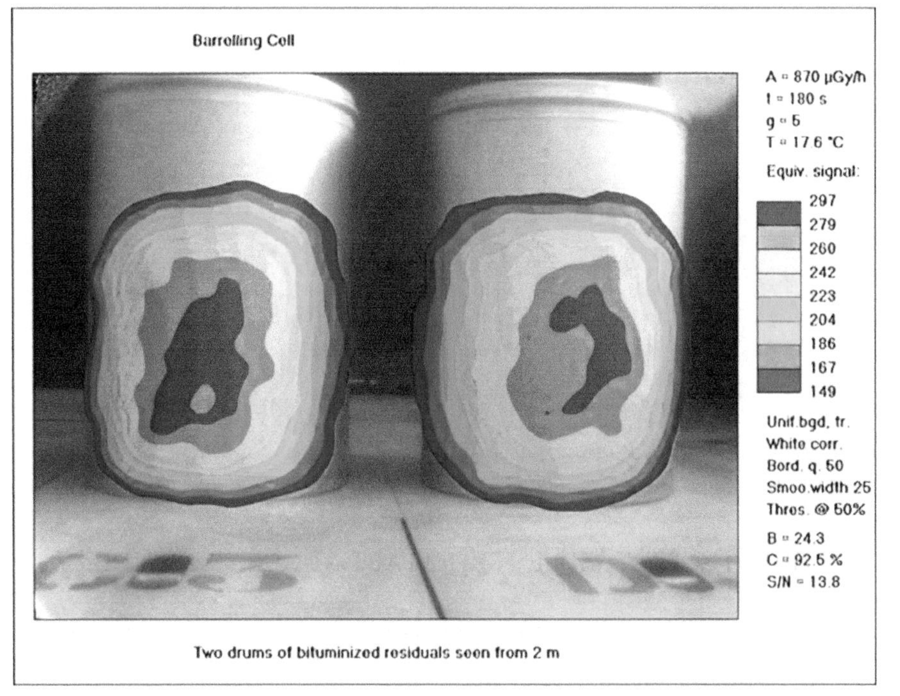

FIG. 12. Example of a gamma imaging system. Courtesy of CEA-LIST.

6.3.7. Destructive analysis

Destructive analytical methods could be required to provide specific and precise radionuclide and chemical composition data for a waste material. Solid samples may need to be dissolved or digested prior to analysis. Isolation or separation of the analyte of interest may be required.

At the Fukushima Daiichi NPP, an organized system of analytical methods for low level radioactive wastes generated from research laboratories had been established before the accident, aiming at simple and rapid determination. The methods were applied to the accident wastes with specific modifications, dependent on the sample. The modified process is shown in Fig. 13. Solid samples, such as rubble and vegetation, were mixed to homogenize after grinding or cutting into small pieces [94]. In case of a need to remove ^{137}Cs, which often increases the background level, ammonium phosphomolybdate was used prior to determination. For analysis of ^{3}H, ^{14}C and ^{129}I, the samples were incinerated to oxidize them to gaseous form and recovered in liquid. Solid extractants were employed to isolate Sr and actinide radionuclides [95].

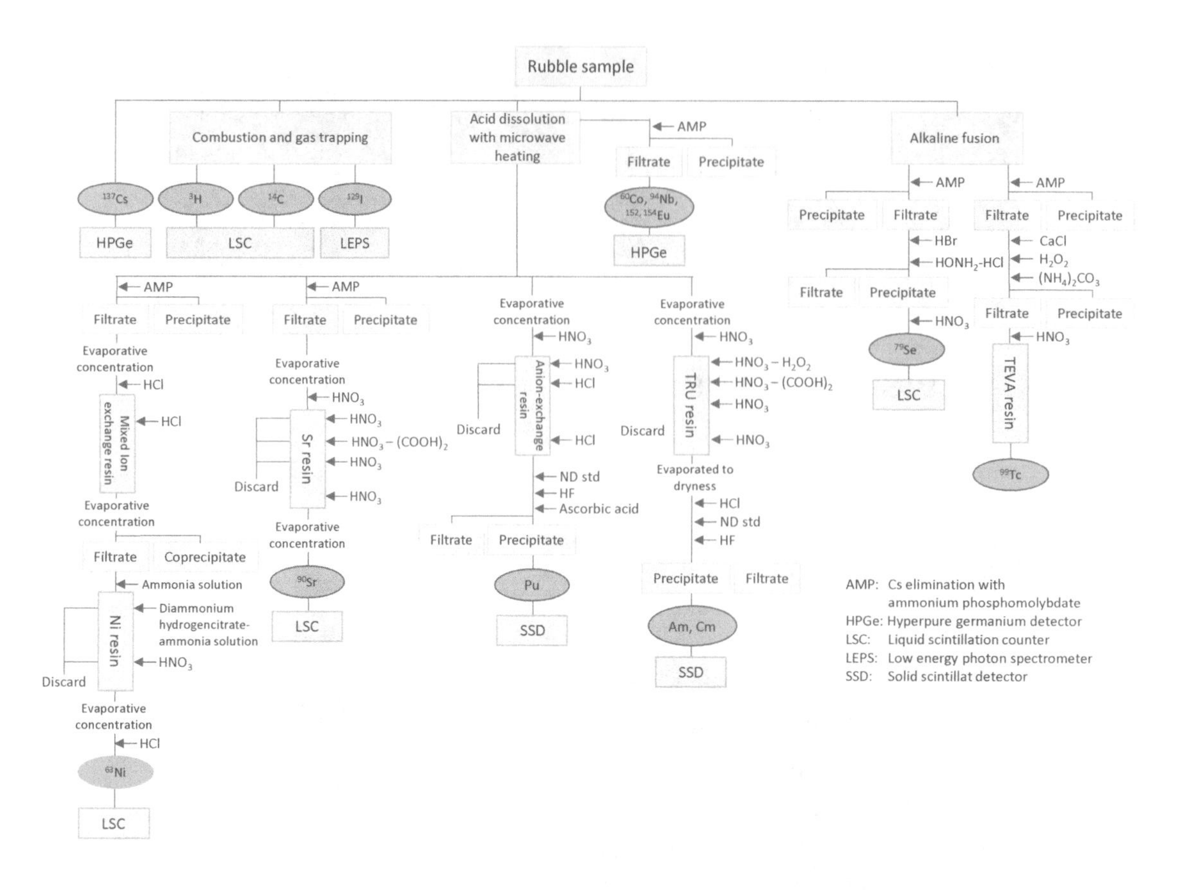

FIG. 13. Analytical method developed for analysis of ^{3}H, ^{14}C, ^{60}Co, ^{63}Ni, ^{79}Se, ^{90}Sr, ^{94}Nb, ^{99}Tc, ^{129}I, ^{137}Cs, ^{152}Eu, ^{154}Eu and alpha particle emitting radionuclides in rubble samples collected at Fukushima Daiichi NPP. Adapted from Ref. [94].

6.4. CHARACTERIZATION EXAMPLES FROM PAST NUCLEAR ACCIDENTS

This section discusses the characterization strategies and techniques employed for, and the lessons learned from, past nuclear accidents.

6.4.1. Three Mile Island accident

Numerous characterization methods were utilized at TMI NPP after the accident to ascertain reactor conditions, determine the chemical and radionuclide compositions of liquid and solid wastes for treatment and disposition, and support decontamination efforts [96]. Routine and specially developed techniques were used to obtain:

(a) Samples of gaseous, liquid or solid material from plant systems, damaged core and surroundings [97];
(b) Chemical and radiological analytical results [98–100];
(c) Visual and dimensional information within the core and the plant [101];
(d) Plant status through calculations and data reduction based on process knowledge.

Reference [96] provides a thorough discussion of the analytical objectives and techniques utilized to obtain the required data. Both successful and unsuccessful data acquisition methods are discussed. The main lessons learned on characterization after TMI NPP were that real characterization and field operations were far more effective than any theoretical approach. Characterization was the bedrock of the technical decision making, planning and engineering. Where characterization was the most difficult, it usually provided the most important data. The characterization effort was an iterative, ongoing process.

A key follow-on task is data management: compiling, organizing and disseminating technical data. Once the cleanup at TMI NPP was underway, sample and analysis planning documents were generated to assist in the organization of sampling and analysis efforts and to compile previous analytical results that were related to current analytical efforts. In this manner, these documents built on previous plans, resulting in an organized structure [102].

As previously mentioned, there was a general feeling of public mistrust after the TMI accident. One attempt to improve transparency and inform the public was through the establishment of a dose monitoring programme within the areas surrounding the TMI NPP that became known as the Citizen Radiation Monitoring Program [103]. The programme was conducted cooperatively by the US DOE, Pennsylvania Department of Environmental Resources, Pennsylvania State University and the US Environmental Protection Agency. The programme provided an avenue for the public to measure radiation dose levels independently and provided a transparent means for the citizenry to evaluate risk as a result of the accident.

6.4.2. Chernobyl accident

A characteristic feature of the Chernobyl accident is the presence of 'hot particles' in the radioactive fallout [3, 104–111] composed of two primary components: particles of finely dispersed nuclear fuel ('fuel particles') and condensation particles, formed as a result of the condensation of volatile fission products (radioisotopes of I, Te, Cs) on the surfaces of different aerosol carriers. Non-oxidized particles were formed during the mechanical destruction of nuclear fuel and are abundant in the western trace of the release, formed immediately following the explosion. In the subsequent reactor fire the nuclear fuel was oxidized, and oxidized particles were released, forming north-easterly and southern traces, as well as superimposing on parts of the initial western trace [106]. Contamination of soils in the 30 km zone forms well marked traces: the narrow western trace of the first release, the north-western trace and the southern trace consisting of the superposition of several traces. The highest soil contamination is associated with the western trace, extending towards the Tolstoy Les village, and the northern trace, extending toward Usiv–Masany villages [108]. The key source of information and data on radiological conditions inside the exclusion zone is the radiation monitoring system, described briefly below.

Continuous monitoring of radiation conditions in the exclusion zone is carried out by Special State Enterprise (SSE) EcoCenter. SSE EcoCenter determines the observation/sampling points, the parameters to be monitored and the sampling frequency. The monitoring programme encompasses the environmental media where radionuclide migration occurs (soils, landscape drainage and biota, air, groundwater and surface water) and is carried out at a regional level, as well as at the specific potentially hazardous technogenic and environmental objects in the exclusion zone. The radiation monitoring network includes 147 observation points and 138 observation wells.

The highest frequency of observations is conducted with the Automated Radiation Monitoring System (ASKRO), which measures external GDR across a network of observation points. Atmosphere, surface water and groundwater, and ecological systems are sampled less frequently compared to ASKRO sampling. Environmental samples are analysed for ^{137}Cs and ^{90}Sr (the main radioactive contaminants of concern), as well as for alpha emitting transuranic radionuclides — ^{238}Pu, $^{239+240}$Pu and ^{241}Am. The yearly monitoring sampling programme typically includes the collection of 4500–5000 environmental samples and conducting ~10 000–11 000 radioactivity analyses.

Methods for radioactive waste characterization are determined by the requirements for the disposal of waste. For example, low and intermediate level, short lived solid and bulk waste designated for disposal at Buryakovka is typically characterized using a Canberra gamma spectrometry In Situ Object Counting System (ISOCS). After characterization, a certificate for disposal is prepared and the waste is transferred to the disposal site. Radioactive wastes that are to be disposed of at near surface storage facilities in the Vektor Complex undergo a characterization procedure using equipment and laboratories located at the Liquid Radwaste Treatment Plant (LRTP) and Industrial Complex for Solid Radwaste Management (ICSRM).

In the immediate and near term aftermath of the accident, numerous short term waste management measures and facilities were implemented within the exclusion zone, which required subsequent in situ characterization. These included:

(a) The creation of nine radioactive waste temporary storage places (RWTSPs) in which contaminated vegetation, topsoil and construction debris were buried on-site in trenches and mounds to reduce radiation fields in the neighbourhood of the NPP.
(b) The creation of three RWDSs to receive radioactive wastes from accident liquidation efforts outside of the Shelter Object: RWDS Pidlisny, RWDS Chornobyl NPP 3rd Stage (also referred to as the Komplexny facility), both of which were closed after the liquidation measures, and RWDS Buryakovka, which is a trench type disposal facility still in operation for the disposal of low level bulk wastes.

Subsequently, in situ field characterization procedures were implemented in 1994–1995 to assess these sites. The characterization methodology had been developed and tested by the All-Russia Development and Scientific Research Institute for Industrial Technology (VNIPIPT) for a 1991–1992 survey in the Red Forest, a 10 km^2 area surrounding the Chornobyl NPP [112–114]. Characterization methods included:

— Visual inspection (e.g. soil subsidence; vegetation anomalies);
— Use of a mobile laboratory equipped for ground investigations and locating trenches and mounds;
— Drilling of 1 m deep survey boreholes on a regular 5 m^2 square grid with gamma logging at a vertical increment of 0.2 m;
— Refinement of the sampling grid when GDR anomalies (> 3 uSv/h) were detected [113].

Based on the results of the characterization campaign, application of the following measurement techniques was recommended to enhance the characterization efforts in the exclusion zone [115].

(1) Reliable portable gamma spectrometers for measurements of angular and spectral distributions;
(2) Passive and active neutron equipment for fissile and fissionable material measurement;
(3) Radiochemical laboratory of the analysis of transuranic (TRU) contaminated samples.

6.4.3. Fukushima Daiichi accident

The characterization of contaminated materials and wastes at Fukushima has involved a range of activities and techniques for organic materials, rubble and water:

(a) Rubble, trees and secondary wastes from water treatment are collected at the site and taken to JAEA (Ibaraki) a few times a year to analyse radioactivity: the total number of samples is ~70 per year. The ratio of ^{90}Sr to ^{137}Cs concentrations in rubble has shown a proportional trend that differs depending on the location where the sample was collected. Continued data collection is underway to improve the precision of the correlation.
(b) It is proving difficult to take samples for analysis from the caesium adsorption vessels and/or sludges generated by the water treatment system, owing to the high dose rates. Consequently, indirect assessment is required, comparing analyses of radioactivity in the contaminated and the treated waters.
(c) Although the amount of data is limited, the inventories of all the waste streams at the site are evaluated based on analyses, estimation and modelling. The analytical inventory estimation is established through creation of radionuclide transfer models based on radionuclide discharge behaviour from a reactor, dissolution behaviour in contaminated water and radionuclide sorption behaviour onto

construction material and adsorbing materials. Inventory data sets are compiled for each waste stream, based on evaluations using the analytical results and modelling.

An important ongoing activity is to increase the number of samples analysed and to improve the accuracy of characterization estimation. JAEA is developing two facilities dedicated to the development of processing and waste disposal technologies. The facilities will comprise two laboratories, with the first focused on low level rubble and secondary wastes from contaminated water treatment and the second facility focused on the HLW produced from contaminated water and fuel debris.

Off-site at the Fukushima Daiichi NPP, the characterization procedure is described in various guidelines, such as the contamination survey guideline [116], the decontamination guidelines [61] and the guideline for the pilot transportation programme [88]. For actual measurement of incoming waste to the interim storage site, new vehicle monitoring systems for air dose rate evaluation are now under development.

7. WASTE COLLECTION, HANDLING AND RETRIEVAL

Key lessons learned:

- Waste can exist in a wide variety of forms (e.g. soil, vegetation, building rubble, equipment, vehicles, putrescible wastes, road cleaning, deposits in rain gutters, ashes, sludges, etc.) and might be dispersed over large geographical areas, including urban areas, forested areas, agricultural areas and remote terrains.
- Waste might exist in both on-site and off-site locations.
- Segregation of wastes by type, material, location and/or radiation levels will provide a basis for optimizing future waste management.
- Remotely operated equipment could be required in areas close to the source of the accident.
- Emergency measures (such as covering some areas with additional soil to provide shielding) could be required in the initial stages of the event. This will have an impact on the amount of material that needs to be treated as waste later.
- Retrieval methods from temporary storage need to be considered (e.g. biological material will decompose and metals will rust during storage).
- Detailed records of stored materials are needed for each location to provide a basis for future retrieval and management.
- Collection and consolidation of contaminated material can reduce the radiological hazard to workers and the public. However, the worker dose and effort involved in collecting and moving material needs to be considered (e.g. it could be safer to leave the material in situ for temporary storage).

7.1. INITIAL WASTE COLLECTION

Many of the challenges in post-accident waste management relate to the initial collection of the wastes, especially those off-site from the accident. The wastes could be widely dispersed geographically and might comprise a wide variety of materials with a wide range of physical, chemical and radiological characteristics. Wastes will include those on the accident site, those created by the cleanup and restoration of the accident site, those created by the restoration of contaminated areas off-site and secondary wastes created by the processing of the other wastes. The wastes at this initial stage will largely be unprocessed (raw) wastes that could have a high degree of loose surface contamination. Collecting the waste can also create additional hazards, such as the creation of airborne radioactive particles (dust) that can easily spread contamination to adjacent areas.

During the initial phases of the accident, contaminated materials may need to be collected quickly, often under difficult conditions, and moved to temporary storage locations in order to reduce the immediate radiation hazards to workers and the public. There is often little time to plan or properly segregate materials by type or characteristics.

Past experience of these initial phases indicates that readily available equipment is used, even if it is not perfectly suited for the task, and difficulties in effectively decontaminating the equipment can be expected. As a result, the equipment itself can become radioactive waste. Such equipment might include construction equipment, readily available transport casks and containers, freight rail cars and similar items.

Existing waste collecting systems can be an effective means for the collection and concentration of contaminants. The following aspects need to be considered:

(a) The concentrated waste forms that can be produced by the movement of contaminated materials through normal treatment processes; for example, water purifying sludge, sewage sludge, incineration ashes, ash in fireplaces, bark, air filters, etc.
(b) Categories and/or classifications of waste during the collection process to ensure that they are segregated effectively for future management.
(c) The durability of collection and temporary storage containers (e.g. plastic bags, metal bins, etc.).
(d) Higher activity wastes, which will need special precautions, such as additional shielding, remote handling, HICs, etc.

Many examples of waste collection methods used after the Fukushima Daiichi accident are described in the decontamination guidelines developed by the Ministry of the Environment of Japan [61].

7.2. HANDLING OF LARGE AND/OR BULK MATERIALS

A nuclear accident can produce significant quantities of large items (such as damaged equipment, contaminated vehicles, construction materials, etc.) or contaminated bulk materials (e.g. soils, vegetation, building rubble, etc.) that need to be collected and managed as waste. These materials can often be collected and moved using standard heavy construction equipment, such as bulldozers, cranes, dump trucks, excavators and similar items. In the case of highly contaminated areas (i.e. with a high ambient dose rate), special adaptation to allow for shielded operator cabs and/or remote control may be required.

If transportation in the public domain is required for the wastes and if the wastes require transportation in a Type B package, then management of those large items becomes more problematic. Type B containers often have small capacities when compared with, for example, ISO Freight Industrial Packages. The time taken to design, license and manufacture large capacity Type B containers can easily be five years or more (see Section 10).

Some examples of adapted equipment used at the Chornobyl NPP are shown in Figs 14–17. In order to perform diverse waste collection and handling tasks in harsh post-accident environments, radical and inventive approaches using available means were required. Basic earth moving and material handling equipment required installation of shielding and modification of existing vehicles for grappling, bulldozing, vacuuming and hauling. This equipment all had either sealed and/or shielded operator cabs. Some equipment could also be operated remotely. Existing military vehicles were adapted in some cases, while other vehicles were custom built. Most of the vehicles ended up as radioactive waste themselves, after they had fulfilled their purpose. A description of the equipment used is available in the Technical Reports Series, published by the IAEA in the 1990s; for example, Refs [48, 117–121].

At the Fukushima Daiichi NPP, cranes and remote controlled robots are used to remove rubble from the reactor and other buildings. Figure 18 shows the progressive removal of rubble at Unit 3, from after the explosion in 2011, through completion of rubble removal in 2014 and the conditions as of 2020. Rubble is transported to the storage area at the site by dump truck.

The disposal of contaminated vegetation is a challenging issue at the Fukushima Daiichi NPP, owing to the volume, areal extent and levels of off-site contamination. Large and, in some cases, remotely operated equipment had been employed to remove vegetation in relatively high air dose rate areas, as shown in Fig. 19.

One of the legacies from the Windscale Piles accident is the contaminated Pile chimney. The 125 m high chimney is surrounded by other legacy facilities holding radioactive wastes and thus conventional demolition approaches could not be used. In 2019 Sellafield Ltd started to demolish the chimney using a 152 m high crane (Fig. 20) to lower 6 t blocks to ground level for disposal as VLLW [122].

7.3. REMOTE EVALUATION OF MATERIALS IN DAMAGED STRUCTURES

Extensively damaged structures can be difficult for personnel to access and may be so highly contaminated that they cannot be entered. Robotic systems have found increasing applications in such situations, where they can be used to explore and identify contaminated materials that will need to be collected. Such systems are currently being developed for specific tasks and there has been significant experience in this area at the Fukushima Daiichi NPP. Inside the reactor building the remote controlled robot shown in Fig. 21 plays an important role in decreasing worker exposure.

Some of the remote controlled equipment being developed by the IRID for use at the Fukushima Daiichi NPP is listed below:

— The MHI-MEISTeR, a remotely controlled robot, which is used to perform decontamination work and concrete core sampling tasks.
— Two types of transformer robots that can modify their posture or shape in water or in narrow spaces surrounded by obstacles. The first two 'shapeshifting' robots that were tested became trapped, but the work continued, resulting in the development of the PMORPH robot. These devices were initially developed to inspect the basement area inside the Unit 1 primary containment vessel as a prelude to the eventual removal of fuel debris.

IRID has continued to develop and deploy further submersible robots for use in the containment structures of the other units at the Fukushima Daiichi NPP to collect data to allow consideration of various decommissioning options and scenarios.

7.4. RETRIEVAL FROM STORAGE

Contaminated materials may need to be collected quickly, often under difficult conditions, and moved to temporary storage locations during the initial phases of the accident. This normally requires that the waste is eventually retrieved from its original interim storage location for processing and eventual disposal.

Since the wastes might not have been packaged prior to their initial post-accident storage, they may be in a degraded condition when the time comes for retrieval. For example, organic materials will be subject to decay, sometimes over very short periods of time. It is advisable therefore, when possible, to package wastes to facilitate subsequent retrieval in a form that minimizes the potential for degradation of either packaging or waste. Corrosion resistant containers can be used with longlasting and robust handling points or designed so that they can be manipulated easily using standard handling machinery. Where degradation of the wastes is possible, best endeavours are warranted to minimize degradation, potentially by conditioning, if the disposal route is well understood. Alternatively, when conditioning cannot be undertaken, rework technologies may be required later.

FIG. 14. Soviet era IMR-2 military tracked clearing engineering vehicle used extensively after the accident for radiological reconnaissance, dismantling, debris removal, waste collection and other cleanup related activities. Courtesy of the Ukrainian National Chornobyl Museum.

FIG. 15. Soviet era military armoured personnel carrier used in the aftermath of the Chernobyl accident to transport workers involved in hazardous cleanup activities and for radiological reconnaissance. Courtesy of the Ukrainian National Chornobyl Museum.

FIG. 16. Modified tractors used in the Chornobyl response. (a) MTZ-80 tractor used for street cleaning to collect radioactive dust and refuse. (b) Heavily shielded modified Cheyabinsk tractor used in cleaning areas adjacent to the damaged Unit 4 reactor and for transportation of supplies. Courtesy of the Ukrainian National Chornobyl Museum.

FIG. 17. Remotely operated earth moving vehicle with a sealed cab used at Chornobyl NPP. Courtesy of the Ukrainian Society for Friendship and Cultural Relations with Foreign Countries.

FIG. 18. Rubble removal and stabilization using cranes at Fukushima Daiichi Unit 3: after the accident — 2011 (top left); during rubble removal — 2015 (top right); current stabilized situation — 2020 (bottom). Upper two photographs courtesy of Tokyo Electric Power Company Holdings.

FIG. 19. Unmanned remote-controlled construction machines used to clear vegetation in off-site areas of the Fukushima region. Courtesy of Kajima Corporation.

FIG. 20. At Sellafield a 152 m crane was installed to demolish the Windscale Pile 1 chimney. Courtesy of the Nuclear Decommissioning Authority.

FIG. 21. Remote control robot (ASTACO-SoRa) used to remove rubble and debris in the reactor buildings, significantly reducing the dose to workers. Courtesy of Tokyo Electric Power Company Holdings.

FIG. 22. ISRWR at the ICSRM, Chornobyl. Courtesy of State Specialized Enterprise Chornobyl NPP.

FIG. 23. Remotely operated dedicated waste container transfer gallery connecting Chornobyl ISRWR to the SRWTP. Courtesy of State Specialized Enterprise Chornobyl NPP.

At Chornobyl NPP, the Installation for Solid Radioactive Waste Recovery (ISRWR) shown in Fig. 22 was constructed at the ICSRM to recover waste from the existing Solid Radioactive Waste Storage Facility (SRWSF). In order to manage the volume and concentration of the solid waste, the ISRWR is outfitted with equipment for remote opening and handling of waste from the SRWSF and its subsequent transfer via a dedicated transport gallery (Fig. 23) to the Solid Radioactive Waste Treatment Plant (SRWTP) for processing and preparation for disposal.

Guidance on retrieving wastes for further treatment has been provided in previous publications, such as Refs [123, 124]. However, the scale of such experience is generally much smaller than that required for a major nuclear accident.

8. PROCESSING OF WASTES

Key lessons learned:

— The waste acceptance criteria of processing or disposal facilities) might not be known with certainty when processing is initiated.
— Additional research and development may be required to qualify suitable treatment and conditioning methods for specific waste types or problematic waste generated as a result of a nuclear accident.
— Construction and operation of new facilities in contaminated areas needs to consider the impact of ambient radiation levels.
— Additional shielding and/or remote handling may be required for some of the waste streams, due to the radioactivity levels.
— Processing could also need to consider non-radiological aspects, such as chemical and biological characteristics, heavy metals and other potentially toxic materials that might be present in the wastes at higher than normal levels.
— Radiation protection criteria will need to be established for any processing performed in uncontrolled areas (e.g. outdoors).

In the context of this publication, 'processing' includes pretreatment (sorting and segregation), treatment, conditioning and packaging of wastes. These activities are discussed sequentially in the following sections of this section. Figures 24 and 25 depict overall schemes for the processing of solid wastes and liquid wastes. The aim of processing liquid waste is either to turn it into a solid form for disposal or to prepare it for discharge.

In a post-accident situation, processing needs to consider short time frames, large volumes and complex, poorly characterized and/or diverse waste streams. The use of fixed or mobile facilities can be considered. Mobile systems are described further in Ref. [62]. Adapting non-radioactive facilities for use in a radioactive waste treatment can also be considered.

8.1. PRETREATMENT

Pretreatment includes activities such as characterization, sorting, segregation and decontamination. Characterization is discussed in detail in Section 6.

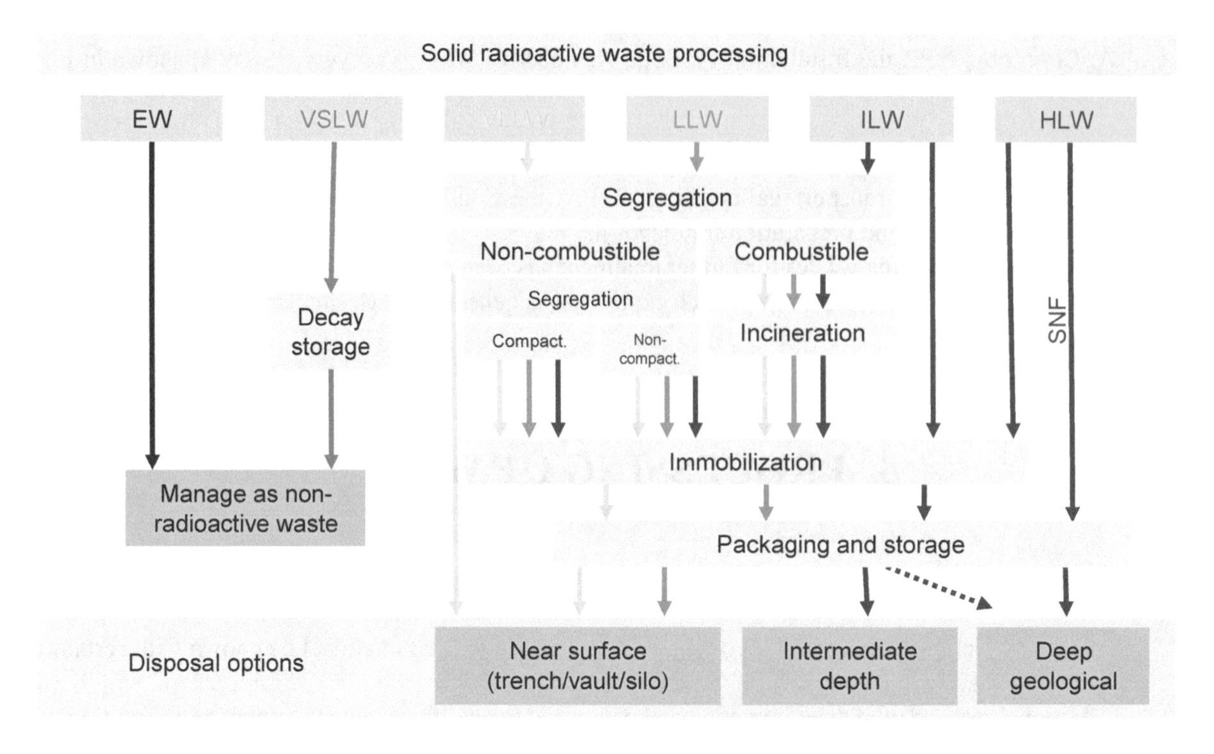

FIG. 24. Schematic representation of solid radioactive waste management. Adapted from Ref. [125].

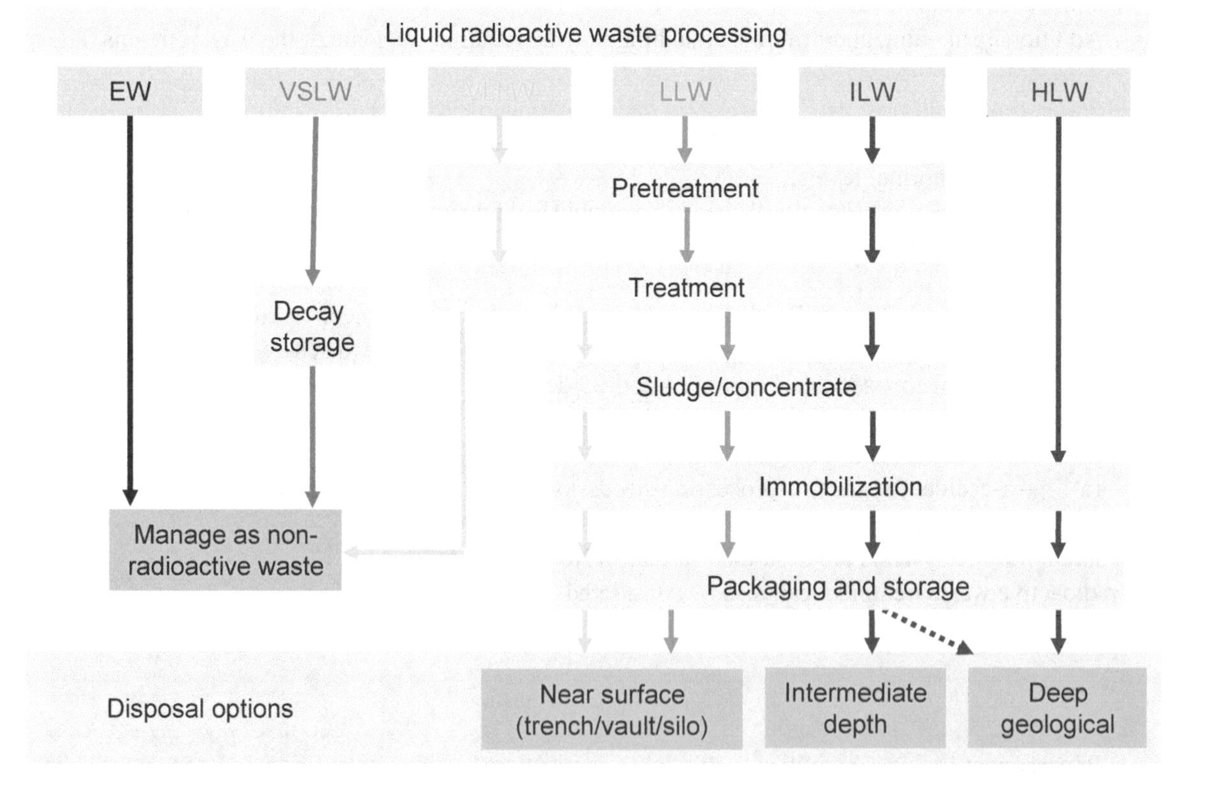

FIG. 25. Schematic representation of liquid radioactive waste management. Adapted from Ref. [125].

8.1.1. Sorting and segregation

In order to facilitate their efficient management, wastes need to be sorted and segregated according to predetermined criteria. These criteria will depend on the plans and available routes for management of the wastes [126]. The criteria can be relatively simple, for example:

— Segregation by materials into combustibles or non-combustibles (e.g. metal);
— Sorting by dose rate.

As the accident recovery and waste management planning progresses, additional sorting and segregation can be utilized:

(a) Level of radioactivity. Activity levels of β–γ emitting radionuclides and α emitting radionuclides. Segregation by activity levels (such as dose rate) is a first priority for protecting workers and the public for the primary waste. Creation of a contamination map (for on- or off-site) is important for segregation of recovery wastes, restoration waste and cleanup waste.
(b) Origin (based on mapping waste generation locations; segregated areas). Classifying wastes according to the reactor unit or point of origin can help with subsequent radiological characterization.
(c) Types of waste. Classifying wastes according to groupings that represent the range of related physical and chemical properties for planning waste treatment, selection of waste container type or for further study of conditioning methods to be implemented in future steps.
(d) Waste generation volume. The volume of primary waste generated is recorded as it is stored; however, it is important to estimate the waste generation volume of recovery waste, restoration waste and cleanup waste to plan improvised and purpose built waste treatment, storage and disposal facilities.

The major factor in segregating and sorting wastes after the Chernobyl accident was their concentrations of radioactivity, or dose rates, but the physical type of waste also featured. For example, the RWDSs were designed to receive solid radioactive wastes from accident liquidation that were not kept in the Shelter Object or stored at RWDS Pidlisny and RWDS Chornobyl NPP Stage III (both of which ceased operation after the liquidation measures) and RWDS Buryakovka, which is a trench type disposal facility that is still operational for low level bulk wastes.

RWDS Pidlisny was designed for the disposal of radioactive waste with GDRs of up to 50 R/h (~500 mSv/h) in module A-1 and 250 R/h (2.5 Sv/h) in module B. The majority of the waste originates from the Industrial Zone of Chornobyl NPP (Promploshadka) as a result of the cleanup activities at the destroyed Unit 4 as well as the damaged Unit 3. RWDS Chornobyl NPP Stage III was designed for the disposal of radioactive waste originating from the cleanup of Units 3 and 4 with GDRs lower than 50 R/h (~500 mSv/h).

Nine RWTSPs were established, located within a 10 km zone around Chornobyl NPP. These contain over 1000 earth trenches and mounds in which contaminated vegetation, topsoil and construction debris from decontamination of fallout in affected villages and the territory of the exclusion zone were buried. All of these sites were closed after the liquidation measures (cleanup) campaign carried out in 1986–1988.

At the Fukushima Daiichi NPP, rubble includes concrete, metal and plastic. Rubble was collected after the hydrogen explosion of Units 1, 3 and 4, and retrieved from the reactor buildings. It was sorted according to dose rate (as indicated in Fig. 26) and stored. Part of the rubble has been covered with soil in order to decrease dose rate. Trees were cut into trunks, branches with leaves, and roots. Part of the tree waste has been chipped and covered with soil to prevent fire and decrease the dose rate. Secondary radioactive waste generated from contaminated water treatment and cooling fuel is currently stored, while slightly contaminated water from sub-drains located around reactors is being discharged into the sea. All of the waste generated in the Fukushima Daiichi NPP is stored on-site.

At TMI NPP, there was essentially no external radioactive material release during and following the accident. Therefore, collection of contaminated construction debris, soils, vegetation, etc. was not an

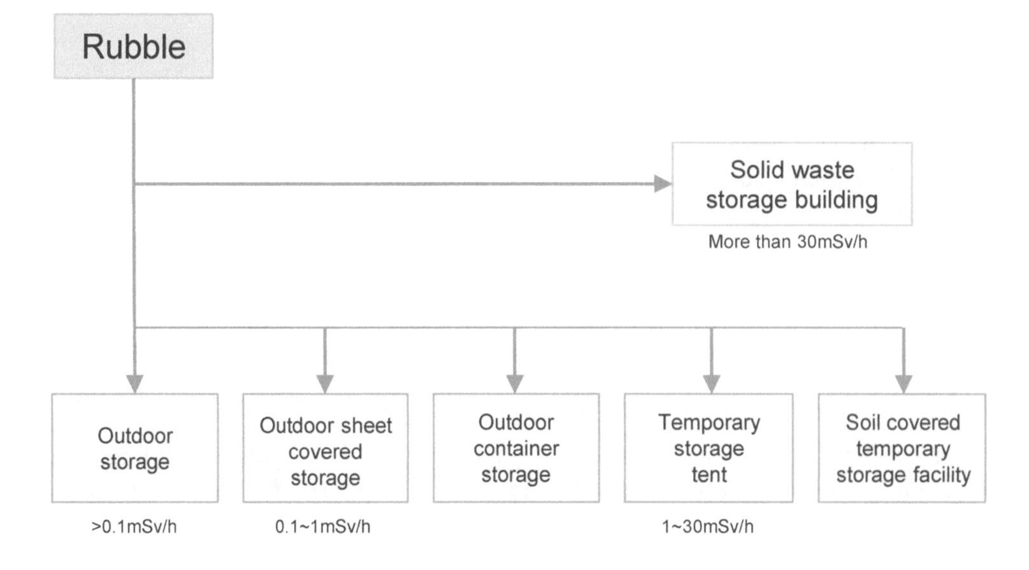

FIG. 26. Waste grouping for Fukushima Daiichi rubble, based on surface dose rate and temporary storage requirements.

immediate problem. Solid waste management primarily involved handling of decontamination equipment and media, water treatment ion exchangers and sorbents, and debris and fuel materials within the plant buildings. However, there was a need for on-site waste staging facilities to support subsequent packaging and transportation of wastes for off-site disposal [1]. Immediately after the accident, a metal shed that was previously used for painting during construction of the plant was used for solid waste collection and temporary storage [1]. In 1980, a programmatic environmental impact statement [30] was written, related to decontamination and disposal of radioactive wastes from the TMI accident, in which the need for new on-site staging and storage facilities was identified.

8.1.1.1. Soil sorting unit

In the event of a nuclear accident, huge amounts of contaminated soil might need to be removed and disposed of. A first step is to characterize the soil radioactivity and to remove soil with an activity higher than a fixed threshold. Soil sorting units generally combine the characterization function (usually counting with large plastic scintillators) with the sorting function.

After the Fukushima Daiichi accident, a high throughput soil sorting unit was developed, recognizing that millions of tonnes of soil would have to be processed [127]. The first assessment of soil contamination around the Fukushima Daiichi NPP and feedback from experience at the Chornobyl NPP has shown that the contamination stays within a depth of 5–10 cm, and contamination concentrates on localized ‘leopard spots’. More than half of the suspected soil was found not to be contaminated. The soils were amenable to handling with an automated soil sorting unit, with a high throughput of up to 150 t/h (nominal throughput 100 t/h). The unit is equipped with large sized plastic scintillators. Its detection limit is ~20 Bq/kg ^{137}Cs (depending on background). Sorting precision is between 30 kg and 40 kg. This mobile unit is packed inside two 12 m ISO containers to allow easy movement from one treatment place to another. The principle of the soil sorting unit is shown in Fig. 27.

8.1.2. Decontamination

Decontamination (e.g. wiping, scouring, or shaving the surface to remove contaminated materials) of some potential wastes (e.g. buildings and equipment) can effectively reduce the volume of waste that needs to be managed and can be considered where appropriate. However, this needs to be balanced by considerations of its effectiveness, the generation of secondary wastes, the generation of airborne particulates and worker dose during the decontamination process.

FIG. 27. Contaminated soil sorting unit used at Fukushima Daiichi NPP. Courtesy of GRS VALTECH.

Decontamination can also be used to allow personnel entry into areas that cannot otherwise be safely accessed, due to high radiation levels, loose contamination or other contamination risks. Standard decontamination techniques can be applied to smaller equipment and buildings. Other techniques, such as those developed for large scale decommissioning projects, can also be applied.

Decontamination after the Chernobyl accident was actively pursued, since the major goal of the cleanup campaign was to create conditions to facilitate continued operation of the three remaining undamaged units of the power plant. Decontamination efforts were concentrated at the site but also in the neighbouring territory. The result of decontamination work was to decrease the contamination and radiation doses by more than an order of magnitude. Details on the decontamination methods and technology utilized are available in the Technical Reports Series, published by the IAEA in the 1990s (e.g. Refs [48, 117–121]).

Decontamination activities primarily rely on field deployable and mobile technologies for application specific cleanup. Experience from TMI NPP showed that wet vacuums, hydrolances, mechanical surface removal and remote controlled techniques were all successful for decontamination of specific areas. Based on this experience, a matrix was prepared to describe the efficacy of the techniques for various decontamination applications (Table 3) [1].

8.2. TREATMENT

Selection of the treatment process depends on the characteristics of the waste. In some cases, no treatment could be one of the options, as was the case with the Chernobyl accident liquidation measures for solid waste. For liquid like, wet solid waste (e.g. sludge), a solidification process is recommended to reduce the potential for migration or dispersion of radionuclides. For example, at the Fukushima Daiichi NPP several means of drying were considered to decrease the water content of sludges (secondary waste generated from contaminated water treatment), to increase the capacity of the solidification matrix.

TABLE 3. DECONTAMINATION TECHNIQUE EFFICACY ASSESSMENT AT TMI NPP [1]

Situation	Low pressure flushing	High pressure flushing	Multiplanar flushing	Reflooding	Fill and leach	Simple tools	Dry/wet vacuuming	Mechanical scrubbing	Hydroscabbling	Hydroscarifying	Scabbling	Scarifying	Steam vacuuming	Dry abrasive blasting	Liquid abrasive blasting	Flex hose nozzle	Strippable coatings	Self-stripping coatings	Chemical foams	Reagents/ detergents
Large quantities of loose, unbounded debris	E[a]	E	E			M[b]	G[c]													
More tightly adherent surface debris removal		E				G		G					G				G	G		G
Contaminated coatings on concrete		M						M	G	G	E	M					G	G		
Uncoated contaminated concrete		M						M	G	G	E	M	G				M	M		
Contaminated coatings on steel		E						G	G	G			G				E	E	M	G
Uncoated contaminated steel		E				M		G	G	G			E				E	E	G	G
Contaminated stainless steel		E				E		G					M						M	G
General area source reduction (rooms/ cubicles)	G	G	E			G	G						E				M	M		G
Externally contaminated vertical surfaces	M	G	G			G			E	E			E				G	G	E	G
Embedded contamination, vertical surfaces				E	E				E	E										
Hollow/porous concrete blockwall contamination				E	E															

TABLE 3. DECONTAMINATION TECHNIQUE EFFICACY ASSESSMENT AT TMI NPP [1] (cont.)

Situation	Low pressure flushing	High pressure flushing	Multiplanar flushing	Reflooding	Fill and leach	Simple tools	Dry/wet vacuuming	Mechanical scrubbing	Hydroscabbling	Hydroscarifying	Scabbling	Scarifying	Steam vacuuming	Dry abrasive blasting	Liquid abrasive blasting	Flex hose nozzle	Strippable coatings	Self-stripping coatings	Chemical foams	Reagents/ detergents
Large area, saturated/embedded contamination				E	E															
Contaminated equipment	G	G				G	G						E	M					M	G
Internal piping contamination																E				
External pipe, cable tray, junction box, etc. contamination						G	G						E						E	

[a] E: excellent.
[b] M: moderate.
[c] G: good.
Note: blank indicates not applicable.

TABLE 4. SUMMARY OF TREATMENT TECHNOLOGIES FOR SOLID RADIOACTIVE WASTES [128-138]

Waste type	Applicable technologies				
	Incineration	Compaction	Decontamination	Segmentation	Composting
	M: technology is mature and widely applied; D: technology under development or limited use; blank: technology is not applicable to this waste stream.				
Combustibles	M[a]	M			
Non-combustibles		M			
Bulk rubble			M		
Metallic solids		M	M	M	
Bulk soils			D[c]		
Incinerator Ash		M			
Vegetation	M	M		M	M

[a] M: technology is mature and widely applied.
[b] n.a.: technology is not applicable to this waste stream.
[c] D: technology is under development or of limited use.

Possible use of municipal solid waste and other existing facilities and infrastructure can be considered after the facilities have been suitably adapted.

8.2.1. Solids

There are several standard techniques for the treatment of solid radioactive wastes, such as incineration, compaction, decontamination, segmentation, etc. These are described in detail in other publications, such as Refs [128–138], and are listed in Table 4. Based on the results of characterization and basic test results on processing techniques, candidate techniques can be narrowed down and issues to be addressed can be identified. While focusing on the narrowed down processing techniques, technical items necessary for studying next steps can also be identified.

8.2.1.1. Treatment of solid wastes at Chornobyl NPP

The waste facilities for processing solid waste at the Chornobyl NPP site and the nearby Vektor Complex to support the decommissioning of Units 1–3, as well as putting Unit 4 into a safe condition, are referred to as the ICSRM, which is equipped with capabilities for the characterization of waste retrieved from storage at the site, with hot cells for segregation and facilities for incineration, super-compaction, cementation and packaging.

8.2.1.2. Treatment of solid wastes at Fukushima the Fukushima Daiichi NPP

At the Fukushima Daiichi NPP the existing waste management infrastructure (incinerator, compactor, etc.) was not used. The facilities are undersized and had previously been repurposed for other uses. Since the accident, two new incineration facilities have been installed on-site. Additionally, off-site incineration facilities generate significant amounts of incineration ashes. Conservatism in the waste management policy in Japan requires that further volume reduction of the ash is pursued, as discussed below.

A flow chart demonstrating the treatment of contaminated soil and waste after the Fukushima Daiichi accident is shown in Fig. 28. After the accident, residues from the incineration of contaminated vegetation (e.g. fallen leaves and attached soil) collected in eastern Japan contained radioactive caesium. The waste was incinerated at a temperature greater than 800°C, with the majority of the caesium being volatilized or vaporized and transported with the stack gases and trapped in cylindrical filters.

Off-site from the Fukushima Daiichi NPP, there is adequate capacity for landfill soil storage. However, for relatively high activity ashes, the capacity of the storage facilities would soon be exhausted, so various technologies have been developed for reducing volume or for decontamination. A simple approach is ash washing, which is effective for incineration fly ash from municipal solid waste, because radiocaesium is contained in water soluble form. Other thermal processes are also available for the removal of radiocaesium, with an efficiency up to 99.9% from materials such as soil, ash and other debris. An ash melting system can also be effective.

8.2.1.3. Treatment of solid waste at Three Mile Island NPP

At TMI, about 75% of the waste generated after the accident was designated as dry activated waste (DAW) [1], resulting in ~3800 m³ of DAW. This category was subdivided into compacted waste (~36%), non-compacted (~60%) waste and stable DAW (~3%). The stable DAW category consisted of wastes that had isotopic concentrations (due to high ^{90}Sr and TRU content), which precluded disposal in a Class A commercial disposal facility. Therefore, it was required that these wastes be placed in High Integrity Containers (HICs; see Section 8.3 for further details) to facilitate disposal in Class B or C disposal systems.

8.2.2. Liquids

Due to their highly mobile nature, it is generally not considered good practice to store contaminated liquids for extended periods of time, and therefore some form of treatment to remove or concentrate and eventually solidify the liquid is normally applied. Basic techniques for radioactive liquid waste treatment are described in a variety of other publications, such as Refs [49, 140–149], and are listed in Table 5. Treatment techniques include filtration, ion exchange (including both general cation–anion methods and ion specific exchange media), evaporation, chemical precipitation and membrane techniques.

The experience gained at TMI NPP was used directly at the Fukushima Daiichi NPP as the basis for several of the liquid treatment systems that were deployed. For example, the KURION caesium removal system employs zeolite as a sorbent; zeolite was used effectively at TMI NPP in the treatment of contaminated water [150]. Organic liquids can also be treated by incineration and other oxidation methods. Small volumes of liquids are often treated by direct immobilization.

8.2.2.1. Treatment of liquid waste at Chornobyl NPP

At Chornobyl NPP the bulk of unprocessed liquid radioactive waste (not including radioactively contaminated waters intended to be processed or organic liquid radioactive waste) is contained in the following storage facilities:

(a) Liquid Radioactive Waste Storage Facility;
(b) Liquid and Solid Radioactive Waste Storage Facility.

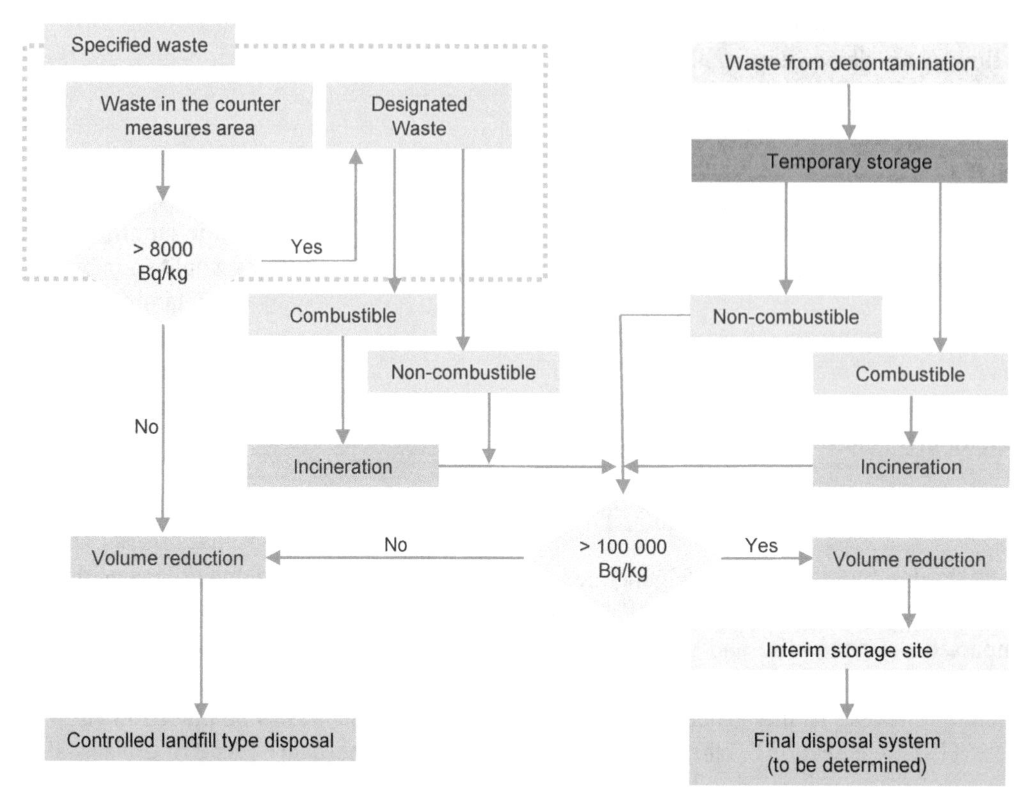

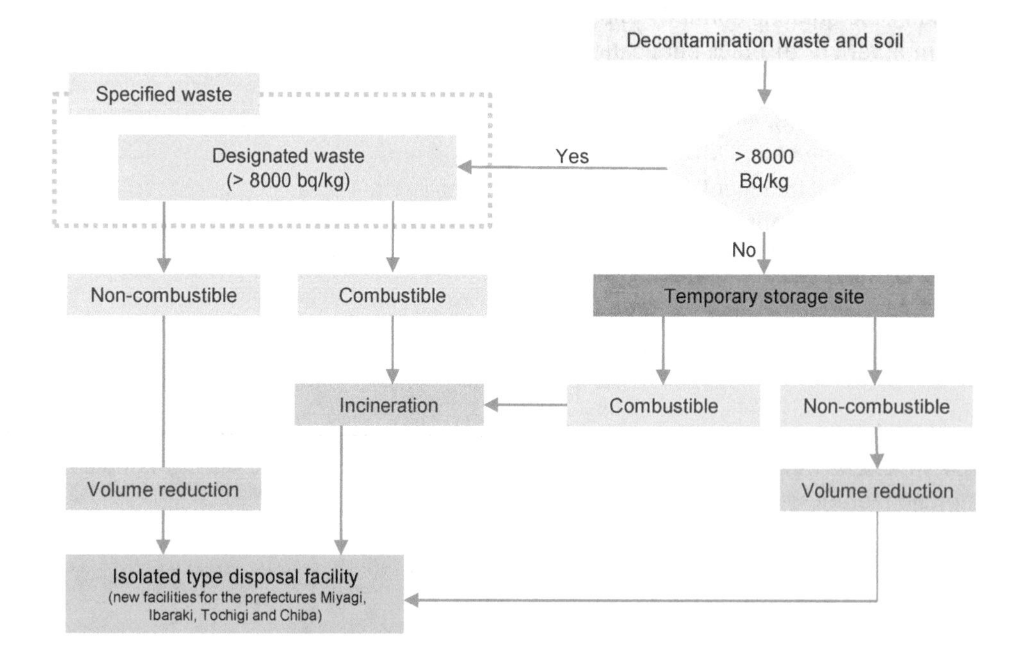

FIG. 28. Flow chart for the treatment of contaminated soil and waste within and outside of Fukushima Prefecture [139].

TABLE 5. SUMMARY OF TREATMENT TECHNOLOGIES FOR LIQUID RADIOACTIVE WASTES [49, 140-149]

Waste type	Applicable technologies				
	Incineration	Filtration/ultra-filtration/ reverse osmosis	Ion exchangea	Evaporation	Chemical precipitation
	M: technology is mature and widely applied; D: technology under development or limited use; blank: technology is not applicable to this waste stream.				
Low activity aqueous		M	M	M	M
Low activity organic	M				
High activity aqueous		M/D	M/D		M/D

[a] List of media in the water treatment system at Fukushima Daiichi:
- Caesium adsorption vessel (KURION) (1.7 × 1017 Bq/vessel (^{137}Cs));
- Zeolite (SMZ, H, AGH, EH, KH);
- Second caesium adsorption vessel (SARRY) (9.0 × 1016 Bq/vessel (^{137}Cs));
- Zeolite (IE-96, IE-911);
- Decontamination device (8.0 × 1014 Bq (^{137}Cs));
- Sludge;
- ALPS;
- Activated carbon, titanate, ferrocyanide material, Ag impregnated active carbon, titanium oxide, chelating resin and resin adsorbent.

These facilities were originally intended to provide storage capacity for operational liquid waste after treatment (i.e. volume reduction) by evaporation, as part of plant operations. The stores were specifically designed for three waste streams:

(1) Evaporator bottom concentrate;
(2) Ion exchange resins;
(3) Pulp (filter perlite).

Currently there is approximately 20 000 m^3 of liquid waste in the storage tanks, of which approximately 13 500 m^3 is evaporator bottom concentrate. The LRTP was constructed in 2006 to treat these wastes. In 2014, active testing of the plant was carried out and 63 packages with immobilized liquid radioactive waste were produced and sent to the Vektor Complex for disposal in 2018. The LRTP became fully operational in mid-2019 and will also treat radioactively contaminated water recovered from the Shelter Object. This water contains a large quantity of transuranic radionuclides, as well as organic substances and other materials used in dust suppression.

8.2.2.2. Treatment of liquid wastes at Fukushima Daiichi NPP

The specific challenges in dealing with accident related liquids at the Fukushima Daiichi NPP include the large volumes involved, complex or varied chemical composition and an inability (or reluctance) to discharge radiologically ‘clean’, treated water. The complexity and variability of the liquid wastes required a multistep process, using different technologies in combination, as shown in Fig. 29.

Absorbers are filled with zeolite to remove caesium, using the KURION system (see Table 5, footnote a. The primary caesium absorbers used in the system include:

(a) Surfactant modified zeolite;
(b) Herschelite;
(c) Silver impregnated engineered herschelite.

In addition, ^{96}IE (zeolite chabazite) and ^{911}IE (crystalline silico-titanate) are used in the second caesium absorber. After use, the caesium absorbents are stored at the site. A mobile version of the water treatment equipment has been developed to process highly contaminated water (Fig. 30).

Following caesium removal, the water is treated further. The ALPS (see Fig. 29) was introduced to improve the treatment capacity of contaminated water. It was difficult for the existing treatment facilities to remove radioactive materials other than caesium, but the ALPS multinuclide removal facility can remove most of the radioactive materials, except tritium. Following treatment by the ALPS, the water is stored in tanks. By the end of 2019, more than one million cubic metres of treated water containing low levels of tritium was being stored in almost a thousand tanks, awaiting a decision by the Government on its final disposal. Discharge to the sea is one of the options considered and is estimated by the MOE to result in minute radiation exposures to the public of less than 0.1 μSv/a — more than four orders of magnitude below natural background exposures [151].

8.2.2.3. Treatment of liquid waste at Three Mile Island NPP

At TMI NPP, over seven million litres of water were processed over the duration of the TMI NPP cleanup [2]. Water treatment and management occurred in four phases:

(a) Phase 1. Stabilization of the situation, transfer of the water to stable locations and treatment of the water when possible.
(b) Phase 2. Large scale water treatment and capture of radionuclides.
(c) Phase 3. Maintenance of the treated water in an acceptable condition for reuse or disposal.
(d) Phase 4. Treatment and disposition of the water for final disposal.

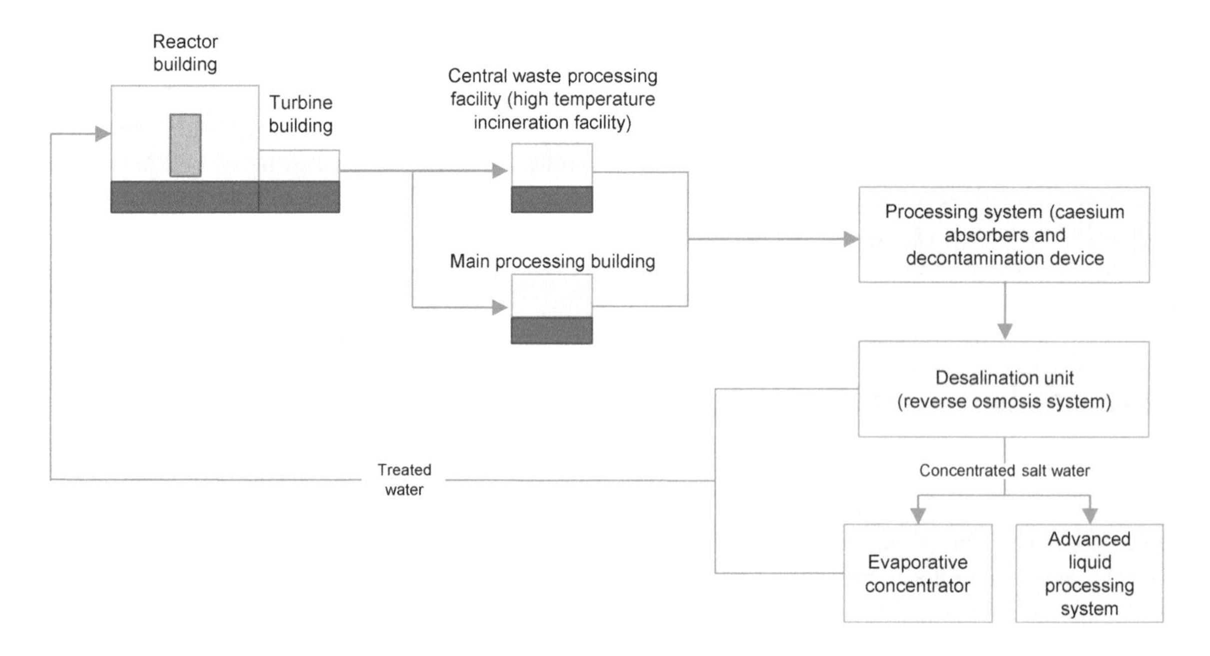

FIG. 29. Schematic of basic contaminated water treatment system at the Fukushima Daiichi NPP.

FIG. 30. Mobile treatment system for contaminated water in Fukushima main trench. Courtesy of Tokyo Electric Power Company Holdings.

Figure 31 provides a timeline for water treatment at TMI NPP. Shortly after the accident, in 1979, water management focused on containing contaminated water and processing/releasing non-accident waste waters. In 1980, water from the auxiliary building was processed. Containment basement water was processed over a two year period from 1981 to 1982. From 1982 to 1985, reactor coolant water was treated. Finally, during the defuelling period (1985–1990), defuelling process waters were treated. Beginning in 1990, stored treated waters were processed using an evaporation process.

Figure 32 provides a schematic of the water processing logic as implemented at TMI. Immediately after the accident, the need for a water treatment system was identified. Within seven days after the accident, the EPICOR I system was installed and placed into operation [1], initially to cleanup water in the Auxiliary Building. The EPICOR I system removed approximately 9×10^{12} Bq (244 Ci) over 1.5 years of operation. It was realized that the EPICOR I system was not capable of treating the large water volumes to be handled, leading to the design and installation of the EPICOR II system. The EPICOR II system was installed in the seismically qualified Chemical Cleaning Building and was designed to facilitate easier insertion and removal of liners containing ion exchange media using a monorail. The building had an existing ventilation system and could provide the required shielding for the highly loaded ion exchanger beds. The EPICOR II system removed approximately 3×10^{15} Bq (80 000 Ci) using organic, zeolite (chabazite) and charcoal media. A schematic of an EPCICOR II vessel is shown in Fig. 33 and the EPICOR II vessels installed in the Chemical Cleaning Building are shown in Fig. 34.

The SDS was installed in Spent Fuel Pool B [1]. The system was designed to treat high level liquid wastes contained in the Reactor Containment Building Basement, the Reactor Coolant System and the Makeup and Purification System that had activities greater than 3.7×10^9 Bq/L (100 μCi/mL). The system was tested extensively at the laboratory scale to optimize design and performance [153]. Water to be treated was molecularly filtered through Zeolite prior to introduction into the resin bed ion exchange columns. The system was operated in several serial and parallel modes utilizing a combination of zeolite (chabazite) and sand media. Following treatment in the SDS, the waters were further treated by EPICOR II. Over 2.6×10^{16} Bq (700 000 Ci) of activity was removed using the SDS system. A cross-sectional view of an SDS canister is shown in Fig. 35.

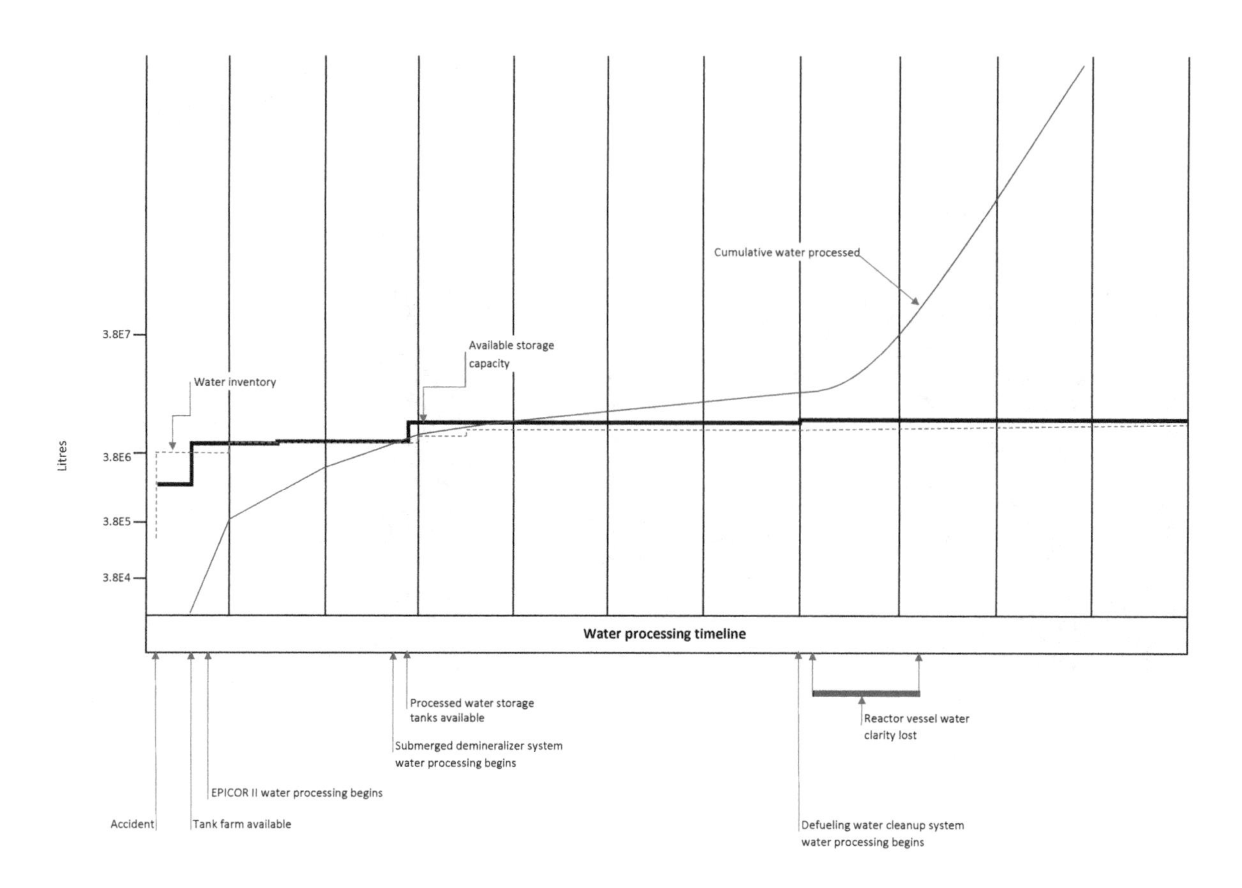

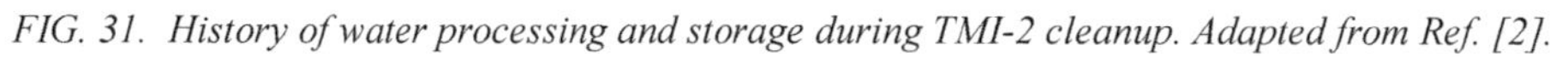

FIG. 31. History of water processing and storage during TMI-2 cleanup. Adapted from Ref. [2].

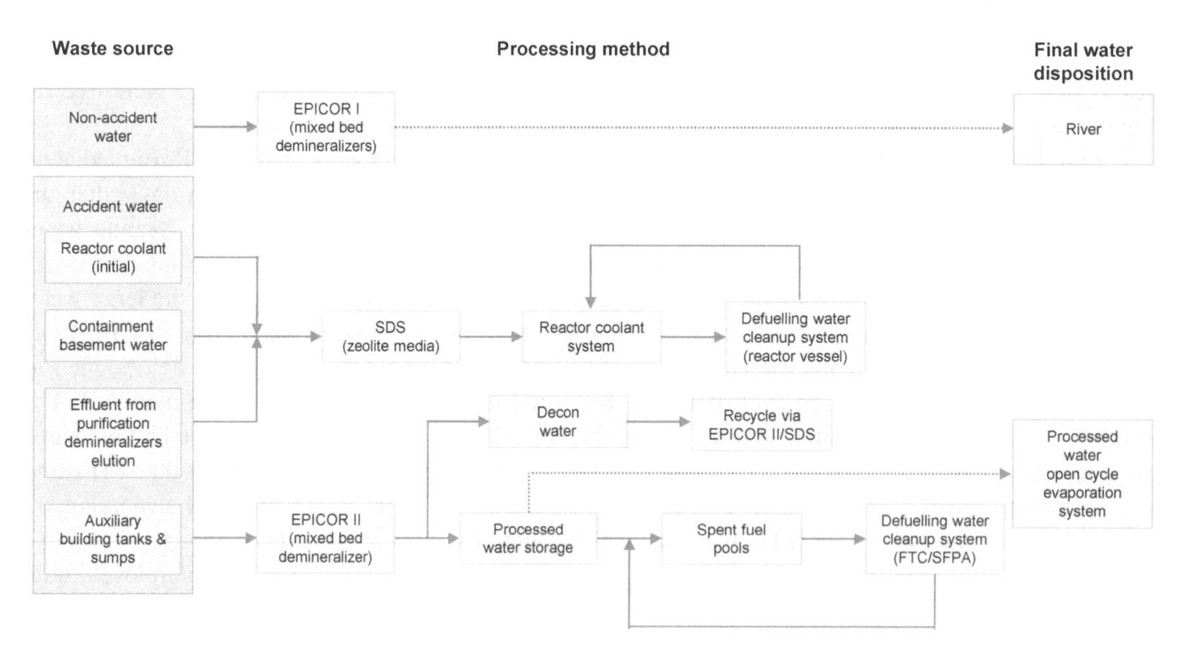

FIG. 32. TMI-2 water processing logic. Adapted from Ref. [2].

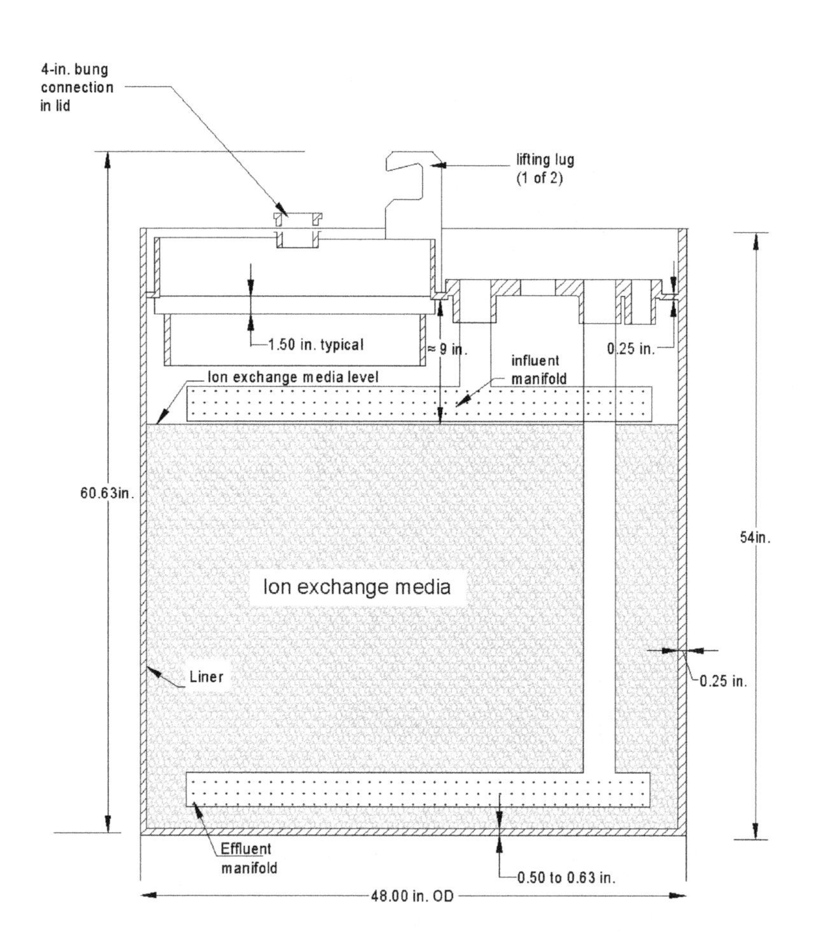

FIG. 33. Cross-sectional view of a prototypic EPICOR II ion exchange vessel. Adapted from Ref. [2].

FIG. 34. EPICOR II vessels installed in the TMI-2 Chemical Cleaning Building. Courtesy of Idaho National Laboratory and the US Nuclear Regulatory Commission [152].

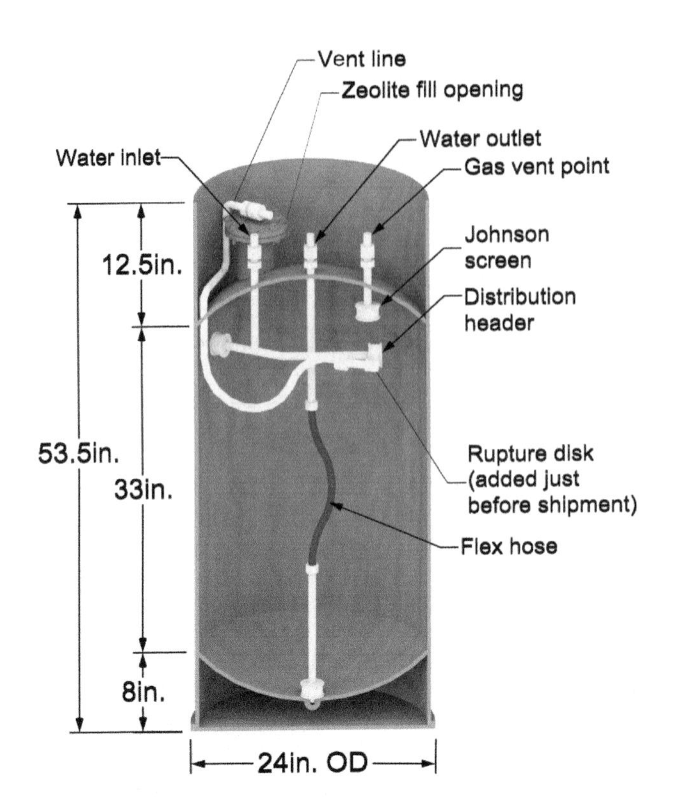

FIG. 35. Cross-sectional view of an SDS vessel at TMI-2. Adapted from Ref. [2].

The Defueling Water Cleanup System was designed to remove solids and maintain Cs and Sr within reasonable levels during defuelling. The zeolite media in the system removed over 2.6×10^{14} Bq (7000 Ci) of activity, while the filtration systems removed particulate matter to maintain water clarity during defuelling.

Disposing of the water after treatment became a significant issue at TMI NPP, primarily due to the remaining tritium concentrations. Although discharge to the river would have resulted in enough dilution that the first downstream user would be exposed to <6% of the regulatory limit, direct discharge was not feasible from a political standpoint at the time [2]. Therefore, stainless steel storage tanks were constructed. By 1986, the amount of accumulated water was ~7.5 million Ls. The concentration of tritium in the water was $\sim 3.7 \times 10^7$ Bq/L (1 μCi/mL) at the time of the accident, but had dropped to $\sim 7.4 \times 10^6$ Bq/L (0.2 μCi/mL) due to mixing of water sources, evaporation and tritium decay. The decision process concerning how best to treat the water considered the following three options:

(a) Discharge to the river;
(b) Evaporation;
(c) Immobilization in concrete slabs.

Although it was assessed to be the most expensive option, evaporation was selected as the disposal method and an evaporator was installed. The evaporation system consisted of two subsystems, one for evaporation and the other for packaging. The concentrated solution was continuously circulated through the concentrator tank. The resulting bottoms from the concentrator tank were sent to a dryer and pelletizer for eventual disposal as LLW to a commercial facility. The Cleanup of Three Mile Island Unit 2 — A Technical History: 1979–1990 [2] provides an account of the decision making processes used to identify and select the water treatment systems.

8.2.3. Gaseous wastes

The basic treatment methods for gaseous wastes involve filtration or scrubbing using a ventilation system followed by atmospheric discharge. To control and improve the accuracy of emission controls, monitoring of the gaseous emissions will be needed at the outlet of an exhaust system. It may be necessary to install adequate ventilation or enhance existing ventilation capacity.

At TMI NPP, $\sim 1.6 \times 10^{15}$ Bq (~43 000 Ci) of ^{85}Kr [42] was estimated to remain in the reactor containment building following the accident. In order to regain access to the building and avoid further uncontrolled Kr releases, the owner and the regulatory evaluated several approaches for its release and developed conservative public dose estimates and a public outreach programme. In June–July 1980, the Kr was released to the atmosphere by a monitored purge during meteorological conditions favourable to dispersion.

The incinerators used for treatment of solid and liquid wastes result in the production of secondary gaseous wastes that also require treatment. Every incinerator in Japan has a flue gas purifying system, such as a bag filter or electrical precipitator (EP). Caesium evaporates as chloride during incineration in some cases, depending on the nature of the wastes, but when rapid cooling is employed to avoid the regeneration of dioxins, caesium condenses and more than 99.9% can be trapped efficiently in bag filters, while more than 99% can be trapped in an EP [154].

8.3. CONDITIONING

Conditioning is defined as those operations that transform radioactive waste into a form that is suitable for handling, transport, storage and/or disposal. Conditioning often includes:

— Immobilization of radionuclides;
— Containerization of waste;
— Additional packaging (overpacking).

Immobilization is defined as the process of converting a waste into a waste form by solidification, embedding or encapsulation, to reduce the potential for migration or dispersion of the radionuclides. For many radioactive waste streams, immobilization takes place concurrently with, or in close conjunction with, conditioning.

Many wastes require further conditioning after treatment prior to long term storage or disposal. There are many standard conditioning techniques, such as cementation, polymer immobilization, vitrification, containerization, etc. These are described in other publications, such as Ref. [125] and references therein.

Conditioning of radioactive waste is an important step to prepare waste for long term storage or disposal. Conditioning can be performed for a variety of reasons, including standardization of practices and/or waste forms, requirements for waste stability based on repository design or safety case criteria, and technical requirements related to waste transportation or otherwise stipulated in regulations or as needed to address societal preferences. In all cases, the reason for conditioning has to be well understood in order to select appropriate methods.

Conditioning provides the link from a raw or partially treated waste to a final form that is suitable for long term storage or disposal. As such, it provides an important safety function in waste management systems and is intimately associated with both the previous steps of treatment and the following steps of storage or disposal.

Some important considerations in selecting and assessing suitable conditioning techniques for accident-related wastes are as follows:

(a) The primary option needs to have a good performance record with respect to radioactive waste conditioning and be known to be both reliable and robust.

(b) Many conditioning techniques, such as cementation, result in significant volume increase. This needs to be carefully considered for large volume accident wastes, such as soil, debris and general rubble. Often simple packaging is a better solution for these waste types, to minimize volume increases.
(c) Large volumes will also set requirements for throughput capacity in order to process the waste in a reasonable time frame. Existing waste management infrastructure often lacks the required capacity and therefore might not be suitable for large scale use after a nuclear accident.
(d) Some R&D may be required to qualify a suitable technique. Many of the waste streams created after an accident are chemically complex, variable in composition, and contain substances that may interfere with the operation of a conditioning process (e.g. organic material, heavy metals, very fine particulate, chlorides, sulphates, reactive metals such as aluminium and magnesium, etc.).
(e) The conditioning technique needs to be applicable to many kinds of waste. For example, a concrete reinforced HIC proved to be acceptable for disposition of resins and other sorbents resulting from water treatment at TMI NPP. The HIC contained a coated corrosion resistant steel liner. Buffering material to control pH was added to the bottom of the HIC. The HIC was closed by bonding the lid to the body using an adhesive gel and flowable grout. A vent system allowed any radiolytic gases generated to escape. A schematic diagram of the concrete reinforced HIC is presented in Fig. 36.
(f) To develop a conditioning technique efficiently it is important to obtain feedback by studying waste form disposability and to identify the requirements for processing from the viewpoint of disposal safety.
(g) Conditioning using in situ techniques is a reasonable consideration [128].
(h) Aspects of techniques that need to be evaluated include:
   - Concept: variation, flexibility, size, capacity, remote, etc.;
   - Feature: merit/demerit, constraint, requirement, problems experienced, etc.;
   - Treatment condition: temperature, pressure, container, containment, etc.;
   - Performance and experiences.

By far the most commonly used immobilization matrix for both solid and liquid waste is cement, owing to its low cost, durability and versatility. Various additives can be used to optimize performance for different types of waste. Cementation equipment is available ranging from very small scale (e.g. litre scale batch sizes) to large, continuous operations (many cubic metres per day). Further details on

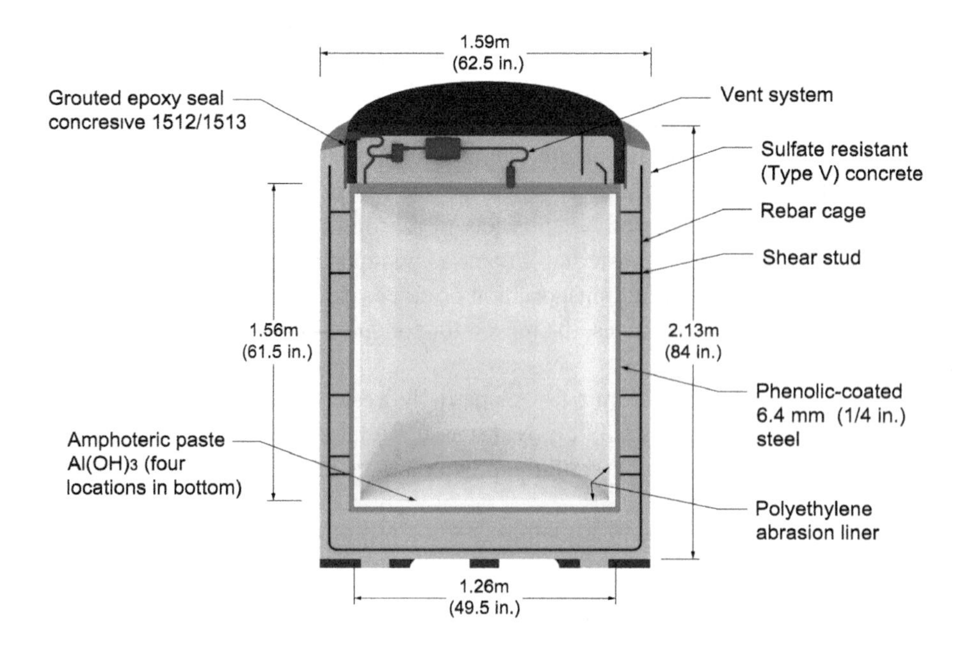

FIG. 36. Schematic showing the design of the concrete reinforced HIC. Adapted from Ref. [2].

immobilization equipment and systems can be found in other references, such as Refs [130, 135, 143, 144, 148, 149].

8.3.1. Solids

Solid wastes can be conditioned either by immobilization or by direct packaging. Immobilization can either be heterogeneous (encapsulation) or homogeneous (mixing of particulate wastes with a binder, such as cement grout). Encapsulation is often used to backfill void space in containers or to surround larger objects (e.g. metal parts, rubble, etc.) in a container, whereas homogeneous immobilization is used for particulates that can be easily 'mixed' into a binder (e.g. ash, ion exchange resins, soils, etc.). Table 6 summarizes the commonly available technologies.

When dealing with large volumes of waste resulting from a nuclear or radiological accident, important aspects to be considered and evaluated include:

— Overall requirements for conditioning, including waste acceptance criteria;
— Available technical options: variation, flexibility, size, capacity, remote, location, fixed versus mobile, centralized versus distributed, cost, etc.;
— Technical features of process: advantages, disadvantages, constraints, regulatory and other requirements, etc.;
— Operating and process conditions: temperature, pressure, container, containment, shielding, etc.;
— Overall system performance and operating experience.

TABLE 6. SUMMARY OF CONDITIONING TECHNOLOGIES FOR SOLID RADIOACTIVE WASTES

Waste type	Applicable technologies				
	Homogeneous cementation	Encapsulation cementation	Vitrification	Metal melting	Polymer immobilization
	M: technology is mature and widely applied; D: technology under development or limited use; blank: technology is not applicable to this waste stream.				
Incinerator ash	M		M		
Non-combustibles	M	M	M		M
Ion exchange media	M				M
Bulk rubble[a]		M	D		
Metallic solids	M	M		M	M
Bulk soils[a]	M		D		

[a] Note that large volumes of low activity bulk materials may not need any conditioning other than packaging in suitable containers (large volume capacity).

Grout forms are widely used for the immobilization of radioactive wastes to facilitate disposal, due to their low cost. Several factors can influence the performance of a grouted waste form, which may be mitigated by grout formulation, process flexibility, package or store features:

(a) Quantity, type and geometry of reactive metals;
(b) Retarding agents such as zinc oxide (ZnO), borates and some organics (although generally used as a retardant, zinc can lead to mechanical strengthening of cements over time);
(c) Accelerating agents such as citric acids or aluminate salts (dependent on initial concentration of citric acid, as it can also retard setting; aluminate salts can also affect dimension stability, particularly for liquid wastes);
(d) Substances that adversely affect cement hydration, such as polyethylene glycol;
(e) Presence of organics that can influence radionuclide mobility;
(f) Presence of organics that can influence reactive material corrosion rates;
(g) Storage environment;
(h) Presence of chloride at low concentrations counteracted by high pH cements;
(i) Presence of organics/free liquors and influence on microbially induced corrosion;
(j) Presence of organics potentially acting as foaming/air entraining agents;
(k) High water content and poor cement powder quality potentially leading to weaker products.

At the Fukushima Daiichi NPP, several solidification experiments with inorganic matrices (e.g. cement, geopolymer, water glass, glass) are being conducted using synthetic material comprising secondary waste generated from contaminated water treatment (e.g. zeolite, sludge). Reuse also reduces storage requirements. Other activities include assembling information on treatment and conditioning technologies for existing waste and presenting candidate conditioning techniques that could be applicable for wastes whose characteristics are better understood (secondary waste from water treatment, rubble, trees, etc.) [155].

The processing facility for solid wastes from the Chornobyl exclusion zone will be equipped with segregation/sorting capabilities and incineration and super-compactor facilities, in addition to cementation/grouting and packaging facilities. However, most of the waste currently buried in trenches and mounds was not preconditioned and can likely be left in place. Wastes contained in the majority of the trenches are expected to have negligible environmental impact over the next 500 years; the time frame envisaged to remove regulatory control from Vektor Complex disposal facilities for LILW. However, a limited number of trenches and mounds with higher activity are planned to be remediated. Waste will be retrieved and either processed or transferred to more suitable locations, including near surface disposal facilities at the Vektor Complex. The HLW and other waste currently buried at RWDS Pidlisny will have to be retrieved and segregated into streams that will be packaged and transferred to deep geological disposal, and streams that can be processed or packaged for near surface disposal. Waste from the RWDS Chornobyl NPP Stage III disposal facility might be handled similarly to wastes from RWDS Pidlisny.

At TMI NPP, solidification of some wastes was required to meet free water requirements and/or stabilized waste requirements prior to disposal [1]. Approximately 100 drums of decontamination solutions were solidified with DOW polymer media. Sump sludge was solidified in a concrete matrix. Also, prior to the development of HICs with dewatering capability, 11 ion exchange liners were solidified using concrete.

Demonstration testing was also conducted to evaluate immobilization of ion exchange materials in polymer and cementitious matrices [156]. Vitrification of zeolite ion exchange media was also demonstrated at a large scale (to vitrify the contents of one liner). However, once disposal was granted for ion exchange material in HICs, it was determined that these immobilization processes were not cost effective [157, 158].

8.3.2. Liquids

Liquids are generally not considered to be suitable for final disposal in a waste repository. Therefore, they need to be treated and converted to a solid form for long term storage and eventual disposal. There are generally two options for the disposal of aqueous liquid waste: discharge of relatively clean material to the environment (e.g. as an effluent) or immobilization into a solid matrix. The former has already been discussed. Typically, liquids are conditioned as a homogeneous mixture (i.e. mixed directly with a binder to form a monolithic solid). Organic liquids are generally treated by incineration or immobilization. This section deals with immobilization/conditioning of liquids into a solid matrix. Table 7 summarizes the commonly available technologies.

As mentioned previously, cementation is the most common immobilization process for liquid wastes. Liquid wastes generated as a result of a nuclear accident can contain constituents that interfere with some typical cementation processes (e.g. organic materials, heavy metals, high chloride content, etc.). Therefore, some research and testing work is typically required to develop a suitable formulation. Such liquids can also be of variable composition, so more than one formulation might be required.

Table 8 summarizes the key features of various cementation processes currently in use. Typically, large scale automated processes are best suited for treating large volume wastes resulting from nuclear accidents.

8.4. CONTAINERS AND PACKAGING

The need for temporary storage of the huge quantities of wastes generated by an accident is one of the key factors that needs to be considered in preplanning exercises. Collection and temporary storage of waste becomes necessary, often very quickly following an accident, to reduce the public and worker radiation exposure due to the spread of radioactive substances. Temporary segregation of these wastes by radioactive levels and type (density and shape) is effective to protect the public and workers from radiation exposure. Selection of optimal containers to retain handling ability and retrievability is also required. The containers will need to provide some protection of the wastes and the containers themselves will need to withstand environmental conditions if they are to be stored outdoors. For example, contact with water will cause organic material to degrade, creating hydrogen and methane, which may pose a fire risk. With higher activity wastes, radiolysis could also be a concern and vented containers might be necessary to allow for the release of radiolytic and other generated gases; for example, see the SDS vessel in Fig. 35 and the HIC in Fig. 36.

The container material thus needs to be compatible with both the waste and the storage environment. Some container materials, such as uncoated carbon steel, might undergo corrosion from both the inside (due to interactions with the waste) and the outside (e.g. from atmospheric moisture). These issues become more important the longer the planned storage duration. Coatings and linings, such as rubber, epoxy and industrial paints, can provide additional protection.

TABLE 7. SUMMARY OF CONDITIONING TECHNOLOGIES FOR LIQUID RADIOACTIVE WASTES

Waste type	Applicable technologies				
	Homogeneous cementation	Vitrification	Crystallization	Polymer immobilization	Geopolymers
	M: technology is mature and widely applied; blank: technology is not applicable to this waste stream.				
Low activity aqueous	M	M	M	M	M
Low activity organic	M				
High activity aqueous	M	M			

TABLE 8. COMPARISON OF CEMENTATION MIXING PROCESSES [159]

Process	Advantages	Disadvantages
In-drum mixing	— Good product homogeneity — Simple processing equipment — Good QC — Suitable for mixed waste streams — Lost paddle variant avoids secondary waste generation and reduces the potential for spreading contamination — Suitable for small waste volumes	— Fixed paddle variant generates secondary waste from paddle cleaning and the potential for spreading contamination — Waste pretreatment may be necessary
Roller mixing	— Simple processing equipment — Suitable for mixed waste streams — Avoids secondary waste generation — Suitable for small waste volumes	— Uncertain product homogeneity — Poor QC — Waste pretreatment may be necessary
Tumble mixing	— Simple processing equipment — Suitable for mixed waste streams — Avoids secondary waste generation — Suitable for small waste volumes	— Uncertain product homogeneity — Poor QC — Waste pretreatment may be necessary
In-line mixing	— Good product homogeneity — Good QC — Suitable for large volume waste streams — Continuous operation	— Secondary waste generation — Waste pretreatment may be necessary — Significant maintenance requirements
Dispersion	— Good for very small volume waste streams — No mixing required	— Uncertain product homogeneity — Waste pretreatment may be necessary

Lower activity wastes can often be collected and stored in flexible bag type containers up to a few cubic metres in volume. The durability of the bag needs to consider that the bag has to be durable enough to last for the full expected time of storage without breaching, and still allow for easy handling and opening when waste is being retrieved. Both of these issues were encountered with the use of bag containers at the Fukushima Daiichi NPP — the initial plastic bags were of poor quality and ruptured after a short time, while the more durable ones used later were difficult to cut open when the waste was retrieved for processing.

Thicker metals or other dense materials, such as concrete, are needed to provide adequate shielding, particularly in provisional or improvised containers used to store relatively high activity LILW. This is because, from the viewpoint of radiation protection, a metallic or concrete container is more effective than a plastic container in protecting from γ radiation emitted by ^{134}Cs and ^{137}Cs in the wastes. Table 9 summarizes the types of containers typically used for solid waste and their applications.

Provisional or improvised containers are in temporary use at the Fukushima Daiichi NPP, pending their treatment or storage in a purpose built facility in the future. ILW (e.g. absorbers with high concentrations of ^{134}Cs and ^{137}Cs) are insulated and stored in a concrete container in box culverts at Fukushima Daiichi. Some contaminated Fukushima wastes were also kept in flexible plastic containers [61]. There are several types with varying durability and water tightness, and selection is predicated upon the waste and storage conditions. Inexpensive containers have often been found to be weak with respect to ultraviolet resistance and generally degrade within a year when used outside.

In some cases, a large capacity container that is readily available and has the capability of holding a large component or quantity of bulk material might be used. These containers can be stacked inside or outside a temporary storage facility, as shown in Figs 37 and 38. This type of open air storage can be applied to store VLLW and care needs to be taken to site such facilities where they will not be affected by flooding and other natural events. Such temporary containers are made according to the specifications for emergency use and they might not be suitable for final disposal. Thus, such containers themselves could become waste once they are emptied. Another solution could be to fabricate disposal overpacks that would allow the waste to remain in the original temporary container. For example, large ISO freight containers are used in some Member States as LLW disposal containers and could be used for such overpacks.

As mentioned previously, regulatory and political issues precluded the disposal of wastes immediately after the TMI NPP accident. Initially, governors in both Washington State and South Carolina rejected the disposal of TMI wastes in the LLW disposal facilities at Hanford, WA and Barnwell, SC. The governor of Washington State allowed wastes to be shipped to Hanford in late 1979, but it was not until 1987 that waste was accepted at the Barnwell, SC burial facility [2]. The subsequent availability of these sites facilitated the disposal of Class A, B and C wastes (i.e. US LLW classification system [29]). Existing approved transportation packages were used for transporting these wastes.

In the US waste management system, (GTCCwastes provided a unique challenge for disposition. An agreement was made between the US NRC and the US DOE for the DOE to accept the GTCC wastes for R&D purposes to evaluate treatment technologies, storage systems, etc. This agreement identified six types of wastes that could be disposed of within the DOE system:

(a) EPICOR II wastes;
(b) SDS wastes;
(c) Reactor fuel;
(d) TRU contaminated wastes;
(e) Makeup and purification system resins and filters;
(f) Other solid radioactive wastes.

TABLE 9. TYPES OF CONTAINER FOR SOLID WASTES RESULTING FROM A MAJOR NUCLEAR ACCIDENT

Type of containers	Example of wastes	Remark
Large metallic container (e.g. ISO freight container, etc.)	Heavy waste, metallic waste, high activity waste, etc.	Container can store large component, high density waste, relatively high activity waste Container usable in outdoor
Box pallet	Metallic waste, concrete, soil, etc.	Container can store large component, high density waste
Drum	Metallic waste, concrete, soil, ash, etc.	Container can store small component, high density waste Mild steel containers may require lining for corrosion protection
Flexible container (e.g. plastic bag)	Concrete, soil, plant, etc.	Container can store waste that does not damage the plastic bag and low density waste Some bags have fibre reinforcing to improve durability
Concrete container	Higher activity waste	Thick walled concrete can provide radiation shielding
Bulk waste (without container)	Concrete, soil, plant, etc.	Temporary stock piling of collected material, pending subsequent disposition Bulk storage of low hazard materials

FIG. 37. Fukushima: temporary waste storage containers. Courtesy of Tokyo Electric Power Company Holdings.

FIG. 38. Fukushima Daiichi NPP: temporary waste storage sites for plastic bags.

The US NRC waived the requirement to solidify the EPICOR II resins in 1981 and efforts were made to develop means to package the resin liners for disposal at DOE sites. The concrete reinforced HIC proved to be acceptable for disposal of the EPICOR II resins at the Hanford site. A schematic diagram of the concrete reinforced HIC used at TMI is shown in Fig. 36.

Extensive evaluation for the SDS wastes was also conducted. These analyses revealed that vacuum dewatering, venting and rendering inert were not adequate for controlling the generation of radiolytic hydrogen. Therefore, a catalytic recombiner system was needed to recombine hydrogen and oxygen back to water. Through subsequent testing, a palladium on alumina catalyst was selected as the recombiner. The resulting solution was to package the liners from SDSs in polyethylene lined HICs and dispose of waste at the Barnwell, SC site, once shipments to the Barnwell site were allowed.

At the Fukushima Daiichi NPP, the slurry and spent adsorbent waste materials generated by the ALPS (see Section 8.2.2.2) that is treating the large volumes of contaminated water are transferred into high integrity containers (HICs; see Fig. 39), which are transported to a temporary storage facility to be stored in box culverts.

At the Chornobyl NPP, two primary waste packages are used. Solid LLW and short lived ILW are conditioned into 3 m³ type KZ-3.0 concrete containers (Fig. 40). Liquid LLW and short lived ILW are treated at the LRTP and conditioned using a cement mixture into 0.2 m³ type MB-0.2IV metal waste drums (Fig. 41). The specifications for these containers are summarized in Table 10.

The KZ-3.0 is a concrete shielded container that provides personnel protection during transport and for storage/disposal. It is licensed for use at Chornobyl (NPP). The container is equipped with two different concrete lids — for transportation and for long term storage or disposal — which are also shown in Fig. 40.

Conditioned waste is transported to the Vektor Complex for disposal in concrete vaults. The disposal strategy calls for the emplacement of the drums in the central portion of the concrete vaults, with the concrete containers placed to form a shielded perimeter around the drums. Following this emplacement strategy, drums in each vault can be stacked in up to seven layers, with the concrete containers stacked accordingly to provide shielding throughout the disposal operation.

Throughout the accident remediation work at the Chornobyl NPP, different waste containers have been used to hold higher level waste transferred to either RWDS Pidlisny or RWDS Chornobyl NPP Stage III disposal sites. More information on the containers can be found elsewhere (e.g. Ref. [117]). However, none of these containers was designed as a long lasting waste package to be used for long term waste containment and immobilization. Waste containment is provided by the heavy structural walls and slabs used at these facilities, which are additionally isolated from the environment by a thick cement cover emplaced over the buried waste. In addition, a multilayer engineered soil cover was designed and installed over both sites to provide additional protection from water infiltration. Bulk wastes were buried in trenches and mounds — containers were not used.

Additional guidance on containers for accident situations can be found in other publications, such as Refs [47, 49, 121, 160].

a.
b.

FIG. 39. HICs used for storing slurry and spent adsorbers from treatment of contaminated water at Fukushima: (a) at the ALPS and (b) at the storage area. Courtesy of Tokyo Electric Power Company

TABLE 10. SUMMARY OF PRIMARY CHORNOBYL WASTE CONTAINERS

Technical data	KZ-3.0	MB-0.2IV (200 L drum)
Design volume (m^3)	3.0	0.2
Outside length (mm)	1940	—
Outside width (mm)	1940	593 (max. OD at lid)
Outside height (mm)	1650	856
Wall thickness (mm)	150	1.5
Empty weight (kg)	5600 (excl. lid)	25
Lid weight (kg)	1000	—
Gross weight (kg)	15000	500
Material	Reinforced concrete	Carbon steel
Use	Solid LILW	Solidified liquid LILW

FIG. 40. KZ-3.0 concrete container used at Chornobyl: lifting and translation configuration (top); transportation lid (middle); storage (bottom). Courtesy of State Specialized Enterprise Chornobyl NPP.

FIG. 41. MB-0.2IV drum used at Chornobyl. Courtesy of State Specialized Enterprise Chornobyl NPP.

9. TRANSPORTATION AND TRANSFER

Key lessons learned:

- Both on-site (transfer) and off-site (transportation) movement need to be considered.
- Transportation of bulk items (soil, rubble, vegetation) might require very large transportation packages. Transportation of large items (contaminated equipment etc.) could also require special consideration. Limitations on transport package size and mass may require package size reduction for transportation of some materials.
- Transportation of bulk liquids might be required.
- High dose rates (both from the waste itself and/or from the ambient environment) could require the use of remotely operated vehicles.
- In emergency situations, strict application of IAEA Transportation Regulations might not be possible: one of the reasons to have an exclusion zone is to eliminate the need for transport through public areas.

The IAEA Safety Standards relating to transportation packaging provide guidance on the safe transport of radioactive materials. The IAEA adheres to these standards in its own operations and Member States apply these standards widely through their own nuclear safety regulations. IAEA Safety Standards Series No. SSR-6 (Rev.1), Regulations for the Safe Transport of Radioactive Material [44] is updated periodically and was most recently updated in 2018 [161]. A graded approach relating to packages and conveyances is employed, depending on the hazard of the radioactive contents. This standard can be used for guidance in the transport of waste materials after a nuclear accident.

An associated IAEA Safety Standard was developed as advisory material for the Safety Standard identified above [44]. This publication aims to further clarify that transport requirements are needed that are commensurate with the inherent hazard of the material being shipped and provides the rationale for this assertion. The publication discusses the importance of packaging and notes that, beginning with the design of the package, adequate safety features need to be considered to the extent practical. By emphasizing reliance on the package, considerations for transportation can be more generalized, minimizing the need for special transport conditions.

IAEA-TECDOC-1728, Regulatory Control for the Safe Transport of Naturally Occurring Radioactive Material (NORM) [43] can provide additional guidance on shipment of low activity bulk waste materials. This publication evaluates exclusions to the Regulations for the Safe Transport of Radioactive Material [44] for the very low activities associated with naturally occurring radioactive materials [43]. The risk based arguments contained in this publication could be applicable to large volume, low activity waste materials (e.g. soils, vegetation, debris) resulting from a nuclear accident.

Transportation of radioactive materials will be needed throughout the waste management programme. The complex and diverse waste streams anticipated after an accident will require a variety of transport casks and transport methods. Waste transportation needs could range from the shipment of very large items to that of small items of varying volumes, with ranges of radioactivity from very low to extremely high. Therefore, in the preplanning phase it is necessary that an organization fully evaluates available shipping methods and routes in preparation for potential future needs. Contingency plans can then be developed that are reflective of the assessments results. For example, for extremely large items, transportation via rail may be the optimal solution, but rail services might not be readily available for a given facility. Contingency planning can also consider options for the development or acquisition of large road ready casks and/or for the identification of viable size reduction methodologies.

Transport on-site (i.e. intra-site transfer) is performed under site transport and package safety cases commensurate with the situation; for example, they can use other protective measures to control dose and impact, such as remote operation, local vehicle shielding, or the distance principle, rather than relying on package qualities. Off-site transport is performed under national or IAEA transport regulations, utilizing licensed packages as required, where safety cases tend to be deterministic. Where wastes are being transported off-site within an exclusion zone or area that has suffered contamination, local transport safety requirements can be considered. Off-site transport to a remote disposal facility is discussed in more detail in Section 11.5.2

The transportation of wastes after an accident will differ significantly from routine operations, with the normally applied requirements being beyond those capabilities in the post-accident condition. Adaptation of transport casks and equipment might be necessary to support the most urgent needs. Reaching out to national and international organizations could be necessary to address packaging and transportation issues.

It is important to ensure that transportation safety regulations are clearly understood, and experts are in place to lead the transportation programme. Dialogue and coordination with regulators could be necessary to develop unique solutions to allow waste shipments or obtain relief from normally applied regulatory requirements in order to mitigate accident consequences.

Some accident waste may require transport in a Type B package (due to the specific activity of the waste or the dose rate from the waste). If no certified Type B container is readily available, then there are likely to be significant cost and schedule implications, even if this only involves adopting a package from elsewhere. For example, the TruPACT-III is licensed and used in the USA for the transport of radioactive waste. However, if a decision is made to use the container in the United Kingdom, additional approval steps would be required, which could take two to three years, following, for example, the procedure outlined by the UK's Low Level Waste Repository Ltd [162]. For wastes with low specific activity, the use of industrial packages may be possible. These can have larger capacity than Type B packages and can be designed, licensed and manufactured much more quickly than Type B containers.

At the Choronbyl NPP, the transportation of waste for burial at RWDS Pidlisny and the Chornobyl Stage 3 facility used heavily shielded vehicles, moving waste either in bulk or in containers. LLW from the site, primarily comprising contaminated soil, is transported for disposal to RWDS Byriakovka in bulk using heavily shielded trucks. The waste from cleanup in the exclusion zone that was buried in trenches or mounds was transported as bulk waste using a variety of shielded trucks. Illustrations of the transportation vehicles used are available in publications such as Ref. [121] and references therein.

Fukushima Daiichi NPP waste has not yet been transported off-site. However, samples of radioactive materials generated at the site are transported off-site to analyse for radioactivity. The radioactive material samples are stored in pails and then loaded onto a truck and transported to the analysis research institute, which is ~100 km from the Fukushima Daiichi NPP. The procedure for transportation is stipulated in legislation. To facilitate transportation, GPS and IT technologies are used, which are also effective for data storage and record keeping.

In the case of TMI NPP, more than 40 000 shipments of radioactive materials occurred over the duration of the cleanup effort. These included shipments of contaminated laundry, laboratory samples and, ultimately, fuel core materials. The shipments were made using rail, highways, air and even US mail. Packaging and transportation became an important part of the cleanup programme and a dedicated packaging and shipping organization was established. The containers used were grouped into three categories:

(a) Commercial 'strong tight';
(b) US Department of Transportation (DOT) specification;
(c) US NRC licensed.

Strong tight containers were used for materials that were classified as less than Class A and were shipped under conditions of 'limited quantity' or radioactive material low specific activity. A specification for these containers was developed that identified the requirements for the type of material being shipped in this type of container. For example, for any waste materials, the container was required to have been constructed of metal with a heavy duty closure.

Type A and B packages fall into the DOT specification container category and were used extensively at TMI. Package compliance for these containers is performance based, meaning that the package (configured as it will be used for shipment) has to meet certain DOT criteria for its intended purpose. The shipper maintains the certification for the package.

US NRC licensed packages pertain to shipments of greater than Class A materials. For these containers, a licence is required to be obtained from the US NRC. For these shipments, selection of the appropriate package for the purpose was important to ensure that compliance with package limits was met.

A large cask was needed to transport the EPICOR II liners contained in HICs. The CNS-14-190 truck cask was identified as the only available option. The CNS-14-190 cask is shown in Fig. 42. A few casks were developed and licensed by the US NRC to facilitate shipments of specific TMI NPP waste materials. The CNS 1-13C Type B shipping cask was developed for transportation of the liners used in the SDSs for disposal (Fig. 43). The cask was qualified for shipments up to 600 W and, over the lifetime of the cleanup effort, transported SDS wastes containing approximately 5.7×10^{15} Bq (155 000 Ci) of activity.

FIG. 42. The CNS 14-190 cask being inspected at TMI. Courtesy of Idaho National Laboratory [25].

FIG. 43. The CNS 1-13C cask readied for shipment at TMI. Courtesy of Idaho National Laboratory and the US Nuclear Regulatory Commission [152].

10. STORAGE

Key lessons learned:

- Both centralized and distributed storage can be considered, depending on the nature and extent of the waste and its geographical distribution.
- It might not be practical to package some wastes (e.g. vehicles, large debris, etc.) prior to storage; however, it may be advisable to package smaller items for ease of handling.
- Reuse or repurposing of existing facilities could shorten time schedules and increase the number of storage options available. Initial temporary storage solutions that can be deployed rapidly could be required in the early phases of an accident.
- Appropriate security measures have to be applied to wastes containing significant quantities of special nuclear material and/or portable high concentration radioactive material that could be diverted for radiological terror purposes.
- All storage solutions have to consider the eventual retrieval of the stored material for processing and eventual disposal.
- Storage will be required for both untreated and treated waste.
- Wastes and storage containers might need to be protected from environmental effects (precipitation, wind, flooding, sunlight, etc.) to prevent degradation of the wastes (especially organic materials) and to control the spread of contamination.
- Complete and accurate records need to be kept regarding the locations and types of waste that are stored.
- Liquid waste storage areas need to be provided with adequate spill containment features.

10.1. INITIAL CONSIDERATIONS

Appropriate technologies need to be selected for storing wastes both on- and off-site. As well as nuclear technologies, a wide range of existing general purpose storage technologies are available and can be considered. Planning will depend on whether the wastes to be stored are liquid or solid, what their activity level is and whether storage is to be on or near the accident site, or off-site. After the Daiichi accident, the MOE developed specific guidelines for storage of off-site wastes [163].

Both short term (temporary) and long term solutions will need to be considered. Temporary or improvised storage refers to a facility that can be put in place quickly to deal with the immediate problems associated with storing collected waste while waiting for the planning and construction of more robust facilities. It is generally only meant for short term storage until a more robust facility is brought into operation. Temporary storage is generally set up in an existing storage facility, building or outdoor space, or in facilities or spaces that can readily be converted to that purpose. Wastes can be stored in simple to robust containers, or in bulk form, without a container, depending on needs. It might be difficult to achieve adequate shielding of wastes in a temporary storage facility. It is also important to implement measures to supress leakage from such facilities and the spread of contamination.

Purpose built storage refers to longer term engineered facilities. Both improvised and purpose built storage facilities can also be used as staging areas to segregate wastes for further treatment or disposal. Although a purpose built facility may be designed for long storage periods, this does not mean that waste will be stored in the facility forever. Storage plans will be affected by planning being carried out for eventual disposal of wastes. The disposal options available or to be developed will affect the times over which storage facilities will need to remain available. Purpose built storage can be constructed either on-site or off-site, or in exclusion zones. They can be constructed in a centralized or a distributed system. They are often designed in a modular fashion to accommodate additional storage needs, as required. These facilities provide storage pending the treatment of wastes for final disposal. However, it is possible that wastes will be stored in this interim state for a significant time, depending on the development and availability of final disposal solutions. Taking this into consideration, it is necessary to consider and plan for the entire waste management life cycle. Considerable time and effort could be required to plan, construct and operate a waste storage facility. It is necessary to consider the human resources required for transporting designing, constructing and operating the storage facility, which will be affected by the siting decision, in particular whether a facility is to be on-site or off-site, and whether storage is to be centralized or decentralized.

Storage planning and implementation will proceed in the following stages:

(a) Estimation/assessment of the condition and amounts of radioactive wastes.
(b) Assessment of the possibility of using existing storage facilities; an indication of priorities and time constraints in dealing with different waste types/categories will become apparent at this stage.
(c) Planning for temporary storage facilities if existing radioactive waste storage facilities do not have sufficient capacity or no suitable storage facility exists. The following options can be considered:
    - Equipment and materials that can be procured quickly (tents, concrete blocks and steel plates for general civil engineering works, etc.);
    - Containers that can be procured quickly (flexible containers, general purpose containers, etc.);
    - Temporary facilities that can be set up (tents, use of idle facilities, etc.);
    - Systems with a short time for licensing (use of simple, proven technologies, etc.); it may be possible to streamline or expedite licensing processes, depending on the regulatory processes in the Member State;
    - Systems with short construction times (land levelling, laying steel plates, covering with soil, etc.);
    - Isolated systems (choosing a remote area, with distance substituting for shielding).
(d) Construction and operation of temporary storage facilities.

(e) Planning for longer term storage and disposal; whether such facilities are required or not will depend upon the ultimate disposal strategy.

10.2. ISSUES IN PLANNING STORAGE

The following principles of radiation protection underpin and need to be applied to the design of storage facilities and the implementation of storage practices:

(a) Containment: avoiding the spread of radioactive substances due to weather such as rain and wind, human or animal activity, etc;
(b) Isolation: reducing the routine radiation exposure from radioactive wastes in storage;
(c) Shielding: reducing the exposure of workers to radioactive wastes in storage;
(d) Time: reducing the radiation exposure time of workers.

For instance, in an accident situation, there is likely to be insufficient time to construct a concrete storage structure that would provide shielding of the waste. The required shielding may thus need to be provided by other means. For example, if a tent is used for temporary storage there might be a requirement to protect the general public by placing the tent in an isolated place, with restricted access. Measures to reduce the radiation exposure of workers would include the setting up of provisional shielding walls (e.g. of concrete blocks used in generic civil engineering works), shortening the work time, clear delineation of the storage area and definition of waste handling, emplacement and monitoring procedures.

An ancillary consideration for both short and long term storage will be the precautionary application of additional security provisions to protect certain categories of material from theft and diversion for nefarious use. Appropriate security measures have to be applied to wastes containing significant quantities of special nuclear material and/or portable high concentration radioactive material that could be diverted for radiological terror purposes. For example, defined de minimis quantities of special nuclear material are generally subject to specific security and material accountability controls that can be applied as soon as practical following an accident. As a corollary, international concerns have been recognized about theft or diversion of radioactive materials in quantities of concern, particularly those that are portable. Prior concepts that the high radiation levels associated with these materials make them self-protecting may no longer be valid, based on the suicidal behaviour of some extreme radicalized individuals. Application of sensible criteria for providing security measures such as barriers and/or surveillance of materials is appropriate. For example, US regulations (10 CFR 37) require such security measures for specific radionuclide concentrations of portable materials (e.g. a category 2 source of ^{60}Co contains $\geq$ 0.3 TBq.).

The following short term issues need to be addressed as storage is being planned:

(a) Location of waste storage facilities: on- or off-site, within or outside contamination exclusion zones, establishing temporary storage arrangements local to contamination hotspots with the intention of later relocating to, for example, decontamination or conditioning and longer term storage.
(b) Type and mode of use of storage facility: temporary facility (e.g. a cask for high level radioactive wastes, a large container, a tent for temporary storage, arrangements for bulk material), temporary use of an existing storage facility etc., waste emplacement methodology and future retrieval.
(c) Required waste storage periods: anticipated period before the availability of a long term storage facility for the wastes and ensuring suitable short/medium term durability.
(d) Immediate treatment of radioactive wastes, as necessary; for example, radioactive decontamination activities and waste volume reduction.

Since the above only relate to temporary provisions, it is important to consider planning and establishing longer term and/or permanent storage solutions, together with any treatment and conditioning facilities, concurrently, to the extent practicable.

While it is recognized that there will be time pressures to ensure that hazardous radioactive materials are contained promptly, nevertheless it is of considerable importance to establish and maintain good records of all wastes identified and under management. It is also important to note the behaviour of the contaminants in the waste and how they may interact with the environment. For example, even relatively low contamination levels can generate a high concentration in leaching water when a huge amount of the waste is stored, the contaminant is soluble and the stored material is not properly protected against water ingress. This may require additional drainage protection (e.g. diverting rainwater and groundwater away from a storage facility to prevent it from becoming contaminated).

Further, it is also important to assess and understand the chemical reactivity issues associated with the waste. An event that occurred in February 2014 at the Waste Isolation Pilot Plant (WIPP) in the USA highlights issues with incompatible materials in stored waste. A small radioactive release occurred through an exhaust air duct due to the rupture of a waste storage drum (Fig. 44) in Panel 7 of the deep geological repository [164]. The investigation into the cause of the event concluded that drum rupture was due to an incompatible mixture of a nitrate based waste with an organic sorbent material used in the packaging process. The reaction might also have been catalyzed by other waste materials present in the drum. To assist in the inspection of the event, a camera system was fabricated that could reach ~25 m into the room, with a swing rotation capability of ~10 m. The system allowed for remote inspection of the ruptured drum and other drums within the room. The primary lesson learned from this event centred on the failure to anticipate and provide pre-job analysis of the hazard created by mixing organic material with the nitrate based salt waste. This led to inadequate evaluation of the drum contents and non-compliance with the WIPP WAC by the waste generator. Further procedural and operational controls were less than adequate. Additional accident investigation findings of on-site conditions included inadequacies in the ventilation system design and safety management programmes, which compounded the effects of the incident [165].

FIG. 44. Ruptured drum located in Panel 7, Room 7 of the WIPP geological repository. Courtesy of the US Department of Energy [164].

At the Fukushima Daiichi NPP in 2015, hydrogen gas was detected in some HICs containing ALPS slurry stored in an on-site storage facility. In addition, puddles of clear liquids had formed on several of the containers, which was seen as a related issue. TEPCO investigated the issues and concluded the likely mechanism to be:

(a) Supernatant solutions were generated by sedimentation of the slurry in the HICs;
(b) Gases were generated by radiolysis of water held in the sediment slurry;
(c) The gases expand the volume of the sediment slurry, raising the level of the solution;
(d) The solution overflows from the HIC and is ejected in some cases through vents and gaps in the lids to accumulate on top of the lids.

TEPCO concluded that the possibility of ignition of the hydrogen gas generated by water radiolysis in the HICs was low. Further, to prevent leakage through the lid, TEPCO drained the accumulated liquid from the HICs in storage and began filling further containers with less slurry, to prevent future leakage and accumulation of fluids on the lids [166].

10.3. SELECTING APPROPRIATE STORAGE SOLUTIONS

The early stages of large scale accident management require prompt strategy decisions regarding the short term management of the radioactive wastes, in order to minimize hazards as far as practicable. Such decisions can include consideration of:

(a) Whether to treat or decontaminate wastes to volume-reduce and/or declassify them, or whether to collect, contain and store them, pending further clarification on management strategy;
(b) Whether to use available storage capability, adapt existing building space, or provide temporary storage facilities;
(c) The timescales for use of temporary facilities;
(d) Selection of container and storage facility type;
(e) Use of centralized storage or distributed storage;
(f) Provision of longer term durable storage arrangements.

Table 11 summarizes the selection requirements and the relevant decision factors.

10.4. IMPROVISED TEMPORARY STORAGE STRUCTURES

Planning and determining the specifications for improvised temporary storage (e.g. tents or containers for solid wastes, tanks for liquid wastes, or their combinations) in order to gain time to store radioactive wastes in a purpose built storage facility requires understanding of basic parameters, such as the generation volumes, activity concentrations and types of wastes. Experience in traditional storage of wastes may not be helpful.

It is necessary to consider the following options to reduce the radiation exposure in the improvised temporary storage of wastes:

(a) Setting up an improvised temporary storage in an isolated, separate area;
(b) Putting such wastes in improvised temporary equipment (e.g. a container or a tent);
(c) Adding a temporary shield (e.g. covering a tank with soil, using a container with shielding capacity, such as a large wall thickness, and employing concrete blocks used in generic civil engineering works);
(d) Use of earth filled roll-off (dumpster) containers as shield walls;
(e) Use of liner baskets within shielded, portable vaults;

(f) Clarifying work steps to reduce the work time for storing wastes;
(g) Lifetime of the storage facility.

It is necessary to keep in mind that this storage is a short term solution pending the planning and implementation of the next step of purpose built interim storage. Various options are summarized in Table 12 and examples are illustrated in Figs 45 and 46.

Temporary storage of contaminated water can present considerable problems if the volumes involved are large, as is the case at the Fukushima Daiichi NPP. For easier construction, a flange connection type tank (Fig. 47, top) was initially used, but leakage from the connections occurred and the tanks were changed to welded types, as shown in Fig. 47 (bottom).

Contaminated water leakage occurred from tanks in which water level meters were not installed and the lowering of the water level was not detected. TEPCO decided to install level meters in the tanks. A double row dyke system constructed around the tanks provides additional protection in case of leaks.

table 11. storage selection process for accident wastes

Requirement	Possibilities	Decision factors
Treatment or storage	Immediate collection and storage, vs. immediate treatment for decontamination/volume reduction/ waste de-classification	— Waste quantities — Availability of cleanup technologies — Availability of storage capacity — Time constraints — Radiation dose rates/contamination levels
Requirement for temporary storage facilities and associated lifetimes	Use of existing storage facilities vs providing temporary storage facilities	— Capacity of existing storage facilities — Waste quantities — Durability of temporary storage facilities — Public acceptability of temporary storage facilities
Selection of temporary storage equipment type	Solid waste: container, tent, etc. Liquid waste: readymade tank, etc.	— Radioactivity level of waste — Location of storage facilities (i.e. isolation) — Waste quantities — Waste treatment/conditioning method — Durability — Shielding and containment capability
Selection of storage type	Centralized storage type vs distributed storage type	— Geographical and social conditions of a location — Storage efficiency — Transportation of waste — Operation and maintenance of storage facilities

TABLE 12. STORAGE OPTIONS FOR WASTE IN IMPROVISED TEMPORARY STORAGE

Target waste	On-site storage	Off-site storage
Solid waste	— Tent (with/without concrete block for shielding) — Container (with/without thick wall for shielding) — Etc.	— Plastic bags (isolate high activity level bags into the centre part or bottom part, with/without covering soil for shielding) — Storage in an already contaminated area
Liquid waste	— Readymade tank (with/without covering soil for shielding) — Reuse of idle storage equipment	— Ready made tank

FIG. 45. Temporary collection sites for contaminated soil and debris at Okuma Town, Fukushima Prefecture. Courtesy of Okuma Town.

FIG. 46. Storage tanks for contaminated water at the Fukushima Daiichi NPP.

FIG. 47. Bulk contaminated water storage tanks at Fukushima: flanged (top) and welded (bottom). Courtesy of Tokyo Electric Power Company Holdings.

10.5. PURPOSE BUILT INTERIM STORAGE STRUCTURES

Pending final treatment and disposal, purpose built interim stores are indispensable to achieve stable and appropriate radiation protection for the public and workers. In planning and determining the specifications for a purpose built interim storage, it is necessary to take into consideration the following options and conditions:

(a) Monitoring, future treatment, transportation, packaging, disposal;
(b) Infrastructure;
(c) Location of storage facility;
(d) Nature and character of waste;
(e) Availability of specialized personnel for design/construction/operation/maintenance;
(f) Responsibilities of planning, construction and operation;
(g) Lifetime of storage facility;
(h) Need for inspection and retrieval;
(i) Inspection requirements for the structure and waste and/or waste packages;
(j) Retrieval of waste and/or waste packages.

The concept of a centralized facility for storing radioactive wastes generated in huge quantities over a wide area is a good option from an economic and practical point of view. In planning such a centralized storage facility, it is also necessary to plan in parallel for both waste treatment and waste transportation to the centralized facility, as well as for eventual disposal.

The concepts for a purpose built storage facility are similar to those described in Ref. [167], although they might be on a larger scale, due to the large volumes of waste to be managed. The applicability of the different technologies available is shown in Table 13.

10.5.1. Purpose built storage at the Fukushima Daiichi NPP

An example of a purpose built facility for temporary storage of caesium absorption towers at the Fukushima Daiichi NPP is shown in Fig. 48.

Caesium absorption vessels are stored in concrete box culverts. The dose rate outside the box culvert is approximately 10 mSv/h. A second type of caesium adsorption vessel, simplified active water retrieve and recovery (SARRY), has shielding, with the dose rate evaluated at approximately 1 mSv/h. High integrity containers (HICs) are stored in concrete box culverts to prevent skyshine and protect from rain and ultraviolet radiation.

10.5.2. Purpose built storage at the Chornobyl NPP

Liquid radioactive waste is stored in specially constructed storage facilities: liquid waste is held in five containers with 5000 m^3 capacity and liquid–solid radioactive waste in nine tanks with 1000 m^3

TABLE 13. TYPICAL APPLICATIONS OF DIFFERENT STORAGE TECHNOLOGIES

Storage type	Unconditioned waste			Conditioned waste	
	Solid	Liquid	Bulk/rubble	LLW	ILW/HLW
Tank		X			
Above ground silo					X
In-ground silo					X
Bunker	X		X	X	X
Above ground shielded vault					X
Trench	X		X	X	X
Shielded building	X			X	X
Unshielded building	X		X	X	X[a]
Mound			X		
Cask					X
Concrete container	X		X	X	X

[a] When used with a shielded container.

FIG. 48. Example of storage of caesium absorption towers at Fukushima. Courtesy of Tokyo Electric Power Company Holdings.

capacity. The liquid waste is treated and conditioned at the LRTP and the product is stored (Fig. 49) pending subsequent transport and disposal in concrete vault, near surface disposal facilities at the Vektor Complex.

Similarly, the ICSRM also provides temporary storage of waste containers (Fig. 50) prior to transport to the near surface disposal facilities at the Vektor Complex. Additionally, temporary storage is provided at the Chornobyl NPP site for packages with HLW and long lived LILW, pending the future availability of a deep geological repository. Wastes are first conditioned into a 165 L drum, and this is then grouted into a 200 L drum.

In addition to the disposal capabilities for short lived LILW, the Vektor Complex will also provide the following storage facilities:

(a) Near surface storage facility for solid LILW with long lived radionuclides (SRW-3);
(b) Near surface storage facility for solid HLW with long lived radionuclides (SRW-4);
(c) Storage facility for vitrified HLW and long lived LILW.

A spent nuclear fuel (SNF) storage facility is being developed at Chornobyl (NPP).

10.5.3. Purpose built storage at TMI NPP

Shortly after the TMI accident, water storage was identified as a significant issue [2]. At the time of the accident, the capacity for water storage was 680 000 L; however, only 190 000 L of space was available. Water management and storage quickly became an operational imperative. The first water storage facility (designated the 'tank farm') was assembled in Spent Fuel Pool A and consisted of six tanks with an available volume of 416 000 L. These tanks were used for storing high activity water. Because direct discharge of treated water to the Susquehanna River was precluded, storage for processed water was needed. Two tanks were constructed to contain treated water that still contained trace quantities of radioactivity. These 1.4 million L epoxy coated, carbon steel tanks were put into service in 1981 to coincide with the ramp up of water treatment associated

FIG. 49. Storage hall for treated liquid at Chornobyl Liquid Radwaste Treatment Plant. Courtesy of State Specialized Enterprise Chornobyl NPP.

FIG. 50. Temporary storage facility for solid wastes at the Industrial Complex for Solid Radwaste Management (ICSRM), showing KZ-3 type containers for storage of reprocessed solid low–medium short lived wastes. Courtesy of State Specialized Enterprise Chornobyl NPP.

with the EPICOR II and SDS. Several other tanks around the plant were also used on occasion. For example, two 850 000 L stainless steel condensate tanks were employed to hold borated water containing low levels of radiation. This water was subsequently recycled and used as flush water prior to treatment.

Very little solid waste storage capacity existed prior to the accident. However, once cleanup commenced, solid waste began to accumulate quickly on the site [1]. These early solid wastes were primarily trash and spent ion exchange materials. As the cleanup ensued, spent ion exchange system prefilters and resin filled liners dominated solid waste activity levels. Due to regulatory and political issues in the aftermath of the accident, shipping the waste off-site for final disposal was problematic. Therefore, temporary storage facilities became necessary. These storage facilities were designed to be flexible to accommodate waste storage and staging for off-site shipping. Immediately after the accident, existing facilities were reconfigured for waste storage. The Paint Storage Shed was used for temporary storage of mildly contaminated trash, protective clothing, decontamination wastes, etc. A temporary storage area for ion exchange media was set up adjacent to the TMI-2 cooling towers within a diked area. The Hot Shop of TAN-607 was initially used for storage of EPICOR II liners in a silo configuration prior to shipment for disposal (Fig. 51). However, the need for an engineered temporary storage facility was quickly realized. A storage facility known as 'waste acres' was put into service. The modular facility consisted of two storage modules with a common drainage sump. The design allowed for units to be added, but additional modules were never needed. Each module contained 6 rows of 10 cells. A module is shown schematically in Fig. 52. The module walls were 1 m thick with a 1 m thick lid to provide the necessary shielding. Each cell was 2 m diameter by 4 m high. A drain line was installed in each cell to drain into the sump. A crane system was used to move ion exchange liners and storage drums in and out of the cells. The Interim Solid Waste Staging Facility was constructed in 1982 to provide additional storage area for LLW drums (208 L) and low activity waste boxes. Again, the design allowed for ready shipment of materials to off-site disposal facilities.

The Waste Handling and Packaging Facility became operational in 1987. This large facility was used to consolidate and package solid wastes and for decontamination. The US $1.7 million cost was justified by significantly reducing waste volumes (25–30% improvements in packing efficiency) and allowed for commercial release of decontaminated materials. Compactible waste was volume-reduced using a stock drum compactor that was installed in the Auxiliary and Fuel Handling Building to further minimize waste volumes.

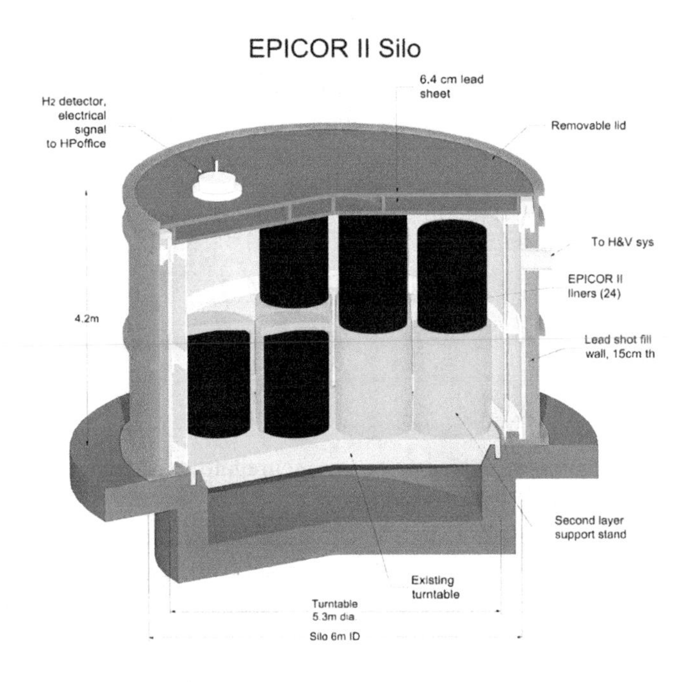

FIG. 51. Schematic of EPICOR II silo in the Hot Shop of TAN-607, TMI. Adapted from Ref. [25].

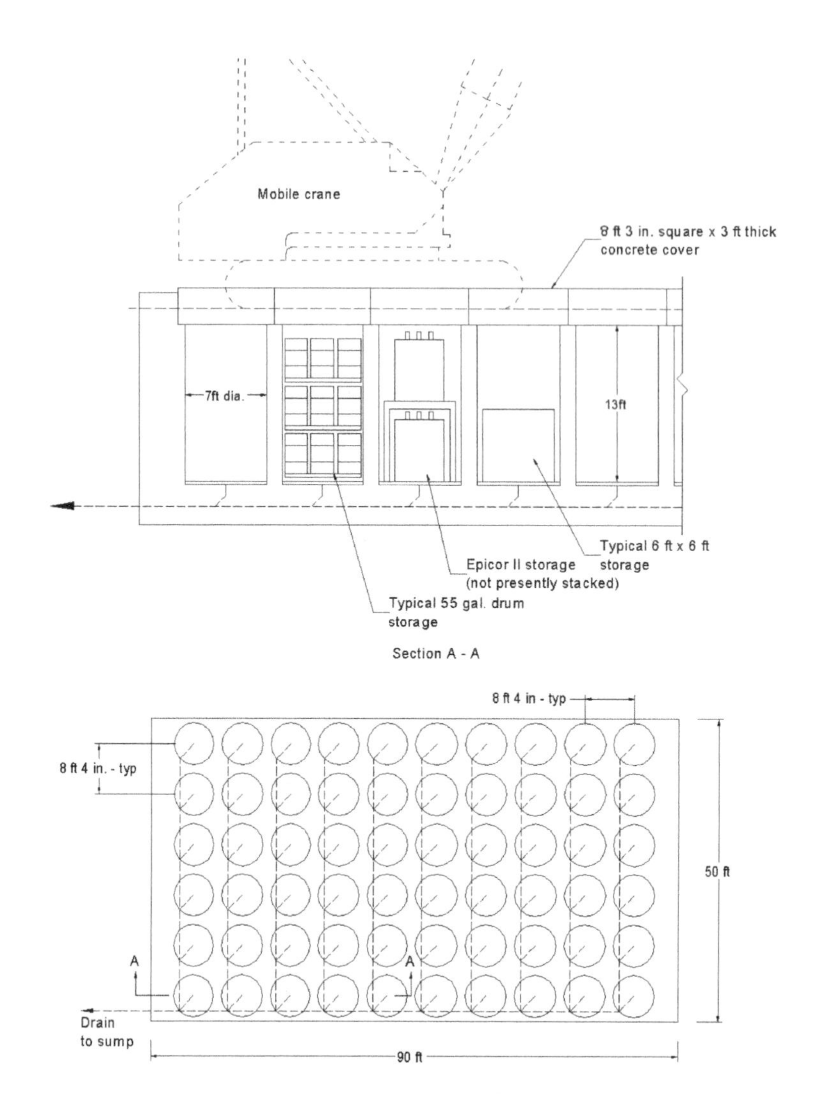

FIG. 52. Schematic of solid waste staging facility at TMI. Adapted from Ref. [2].

11. DISPOSAL

Key lessons learned:

- — Failure to inventory, document, characterize and segregate wastes adequately has limited future disposal options and substantially increased costs.
- — Rapid, unplanned burial of unsegregated, uncharacterized waste has required resource intensive recovery and reburial at substantial cost, with significant worker exposures.
- — Non-radiological constituents present in waste have increased the technical complexity and cost of disposal.
- — Identifying disposal designs based on experience with existing facilities can expedite the development of new disposal capacity for accident wastes.
- — Timely involvement and transparent information sharing with communities and other stakeholders is an important part of disposal planning and implementation.

— Siting of new disposal facilities has been controversial, regardless of whether the proposed site is near to or distant from the location of the accident.
— Diluting waste in the vicinity of the accident by ploughing under soil has increased total waste volumes and related disposal costs.
— Temporary staging or storage of waste or using liquids to decontaminate vehicles, equipment, or other material under inadequate containment conditions have contaminated soils and groundwater, increasing waste volumes and disposal costs.
— Use of very large capacity disposal units, constructed in phases, has provided economies of scale that have substantially reduced unit and total disposal costs.
— Disposing of waste inconsistently with the safety case (e.g. in environmentally unsuitable locations or without adequate engineering barriers) has led to the early closure of disposal facilities and future waste recovery and redisposal at a substantial cost.
— In past major emergencies, available funds have proven insufficient for disposal, at times requiring international funding. Such support has generally focused on priorities other than disposal.

The wastes generated in accident response, cleanup and remediation will require eventual disposal, on- or off-site. Depending on the level of national preplanning already accomplished, it might be possible to dispose of some wastes permanently during early remediation work. The disposal of remaining wastes will require the development of new facilities — a process that could be expedited if adequate preplanning is accomplished (see Section 4). Decisions on the disposal of remaining wastes might need to be postponed if on-site disposal is not appropriate and there is not a comprehensive national waste management strategy in place. For example, the most appropriate disposal solution for some wastes could involve using a multiuse national facility that might not be available for some decades. The latter situation would arise in almost all Member States with respect to routing some categories of accident wastes to a national geological disposal facility (GDF) for higher activity wastes. Thus, storage and disposal planning need to be closely integrated. In general, it can be noted that the disposal options for accident waste follow the same strategy as for disposal of wastes from normal operations of facilities. Disposal options for all waste classes are discussed in Design Principles and Approaches for Radioactive Waste Repositories [26]. Figure 53 summarizes the design concepts for waste disposal.

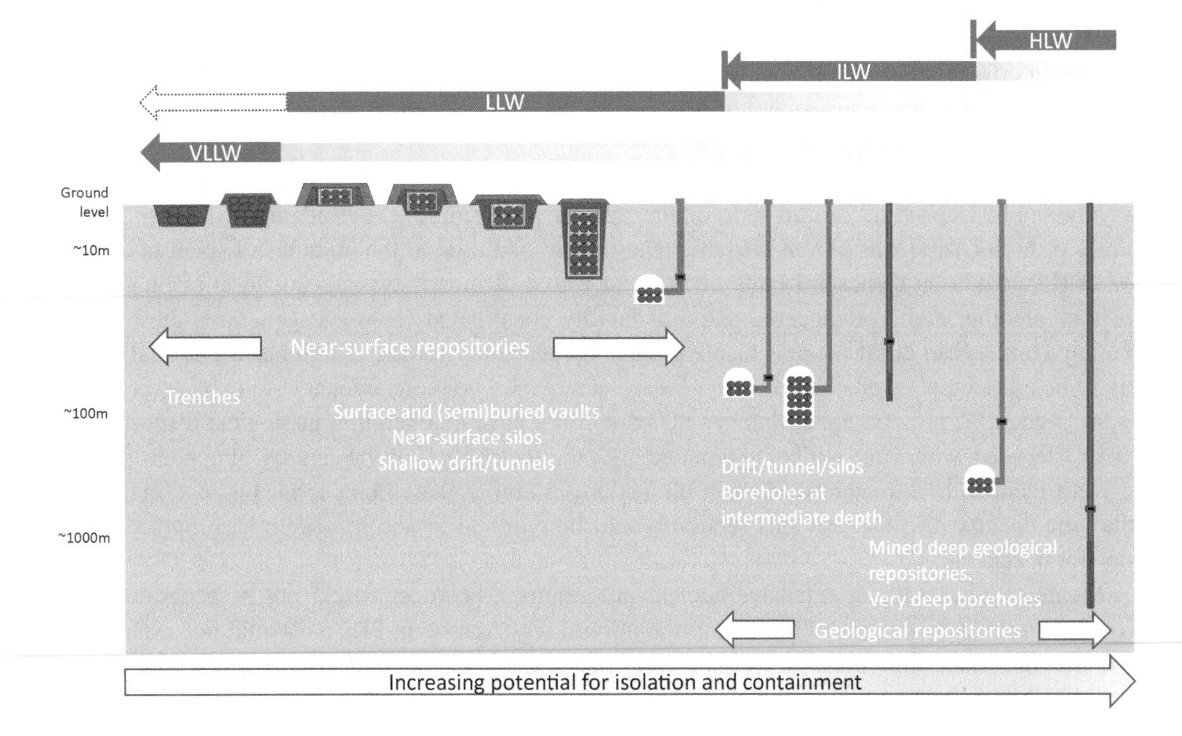

FIG. 53. Schematic illustration of the range of disposal options, from surface to deep, currently considered or implemented for different classes of radioactive waste. Reproduced from Ref. [26].

Several questions arise when developing and implementing a disposal strategy. These include:

— How might waste volumes impact on national waste disposal activities?
— What types of disposal facility might be required for the different types of waste?
— How much disposal volume might be needed?
— How can sites be found for disposal facilities?
— How can the operation of a disposal facility for accident wastes be optimized?

These general issues are examined sequentially in the following sections in this section.

11.1. WASTE VOLUME IMPACTS ON NATIONAL WASTE DISPOSAL ACTIVITIES

Identification of appropriate disposal solutions is based on the hazard represented by the different types of waste being managed. In the context of the following discussion, 'hazard' is used as a general term that encompasses radionuclide content (activity) and persistence (half-life), as well as chemical or biological components.

Figure 54 depicts the general expected relationship between volumes and hazard for wastes generated from a nuclear or radiological emergency, together with that for routine non-emergency operations. It is assumed that the emergency waste is predominantly high volume, low hazard material and could be orders of magnitude larger in volume than normal operational wastes, as has been seen following the Fukushima Daiichi accident. In this situation, it is highly likely that there will be insufficient existing disposal capacity to accommodate this large incremental volume of waste, which is the situation primarily addressed by this publication. However, even if it could be partially or totally accommodated, decision makers may conclude that the best course of action is to construct new capacity regardless, to avoid displacing capacity that is relied upon for the disposal of wastes generated from routine operations. Thus, one potentially unquantifiable strategic factor to consider is the disruption to the normal national waste management system and the impact on other waste generators who depend on the overall waste management infrastructure. In the case of an emergency, parts of this infrastructure, including capabilities typically dedicated to the management of non-radioactive waste, could become an important component of an overall response strategy.

By contrast, the volume of high hazard waste resulting from the emergency would hopefully be relatively small. Depending on the state of the nuclear programme in the affected Member State, the volumes of high hazard waste from the emergency might add little to the volumes of spent nuclear fuel, HLW or ILW requiring disposal. In this situation, it would be more advantageous to utilize a geological repository or other highly engineered disposal facility constructed for waste generated during routine operations, rather than construct new disposal capacity. A facility for these high hazard normal wastes is likely to be a highly planned, one of a kind facility and would be more amenable to expansion by 5% or 10%, as needed. At present, most Member States will not have an operating geological disposal facility for many decades, with some not being planned until the second half of the century. It appears inevitable that, were there to be a major accident in almost any Member State, long term, purpose built storage of the type discussed in the previous section would be required to accommodate high hazard accident wastes for decades.

Member States without extensive nuclear programmes, however, might not be generating higher activity waste under normal conditions (i.e. the normal waste curve in Fig. 54 would not extend all the way to the right side). This situation would create the additional dilemma of disposing of waste of greater hazard than was foreseen by the national waste management programme.

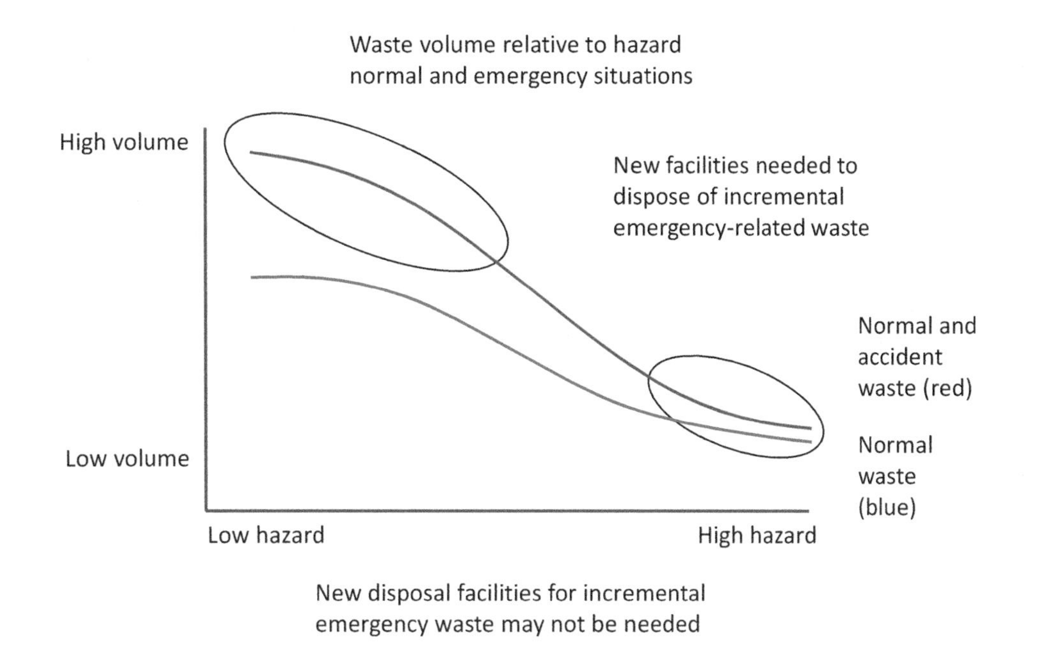

FIG. 54. Relative proportions of comparably hazardous normal and accident wastes that might ultimately require disposal.

11.2. IDENTIFYING THE TYPES OF DISPOSAL FACILITY THAT MIGHT BE REQUIRED USING A GRADED APPROACH

Decisions on appropriate disposal technology selection can benefit from a risk based, cost informed, graded approach. The following principles are suggested to implement such an approach:

(a) Wastes are disposed of using the simplest disposal concept available for which safety and environmental protection can be demonstrated, consistent with the hazards present. If no suitable disposal route exists, then the wastes concerned will need to be stored until a disposal solution is made available.
(b) Disposal of the most hazardous wastes will require the use of greater levels of engineering to provide adequate isolation and containment, including disposal at greater depths.
(c) Assuming that disposal facilities are available, consideration can first be given to if and how they could be used, prior to considering developing new facilities. Additional regulatory authorizations and therefore additional analyses may be needed for existing facilities to accept accident wastes, in order to address the requirements already imposed by existing WAC and operational procedures.
(d) Wastes are segregated so that slightly contaminated, shorter lived waste constituents are not combined with highly contaminated, long lived constituents (see Section 8.1). Segregation by radiological and non-radiological hazard is also desirable.
(e) Adopting clear criteria to determine what material has to be handled as radioactive waste and what can be cleared for disposal as non-radiological waste, or for recycling or reuse.

Table 14 illustrates the graded approach and indicates what type of disposal technology could be most suitable for certain types of emergency wastes. This table is only qualitative, and decisions on waste disposal will need to be made on a case by case basis, taking into account actual accident conditions, waste characteristics, applicable regulatory requirements and the overall accident cleanup strategy that is adopted.

Although an emergency situation could result in very large volumes of waste that will likely require the development of new disposal capacity, disposal practices will need to be conducted in as 'normal' a manner as possible, to ensure operational and long term safety. To that end, a graded approach can be effective when the level of control to be exercised over emergency generated waste is commensurate with the hazard of the material, such that the level of safety achieved by permanent disposal is comparable to that in a 'normal' operating situation.

Preplanning (see Section 4.1) can identify existing facilities that could potentially be employed and could develop generic disposal approaches and preliminary designs to facilitate the implementation of new facilities if the need arises. However, the final demands on these facilities will depend on the nature of the emergency and the volume and characteristics of the waste. Similarly, the safety case for the facilities will be constructed much as it would be for a 'normal' case, with the same elements and endpoints. Preplanning would help to facilitate safety case development and the development of preliminary WAC. Generic WAC can be identified during a preplanning exercise that are appropriate to the types of facilities that might be required, based on international experience with operating similar repositories and the safety cases that have been produced for them. Knowledge of the intended disposal routings that will become available and typical WAC for the intended facilities would permit the initial stages of waste management to proceed efficiently.

11.3. NUMBER, SIZE AND TYPE OF DISPOSAL FACILITIES NEEDED

The physical space occupied by disposal units, roads, parking areas, waste receiving and inspection areas, and administrative and other support functions, as well as an undeveloped buffer zone for monitoring and maintaining distance from surrounding land uses, is also an important consideration. Space needs could increase if rail access or waste processing operations are intended. In keeping with economy of scale benefits, reserving space for future expansion of disposal areas could also be appropriate.

Early emergency volume estimates might also grow significantly over time, for a variety of reasons. This has to be taken into account when determining the size of the site that will be needed, including adequate space for the disposal units and a surrounding undeveloped area. This undeveloped area may be used for environmental monitoring; the construction of required roads; administrative buildings; waste receiving and unloading facilities; and potential waste testing, processing, packaging and other activities.

Large volumes of LLW or VLLW may be cost effectively shipped to large capacity disposal facilities developed specifically for such waste. Substantial experience exists in disposing of very large volumes of building and other structural demolition debris, municipal solid waste, hazardous chemical waste, uranium mill tailings and other mining waste, and low or very low level radioactive waste, in a manner that is protective of public health and the environment. Much of this experience involves the disposal of bulk, unpackaged waste using near surface disposal design concepts. While level topography is commonly selected, experience also exists in disposing of waste in valleys where there is limited land availability. Smaller capacity facilities that cost more to construct and operate with a higher unit disposal cost could then be used solely for lower volumes of higher activity wastes.

The experience in France with the Centre de l'Aube and Morvilliers disposal facilities demonstrates that different disposal facilities handling different waste types can be located in close proximity [76]. This can allow the substantial cost of common roads, utilities and administrative support infrastructure to be shared. Allowing sufficient distance for independent facility performance monitoring to take place is also an important consideration.

Depth of disposal and potential use of engineered barriers to limit contaminant migration varies, based on waste type, local soil, hydrological and other site environmental conditions, as well as climate, proximate land uses, applicable regulations and other factors. Shallower burial depths might be needed to provide adequate separation from groundwater or permeable bedrock. The use of compacted clay liners, synthetic membrane liners or combinations of these barriers is common to prevent migration of liquids out of burial units. In the case of very low permeability on-site soil conditions, or desert areas with very low precipitation and high evapotranspiration, liners might not be required.

TABLE 14. GRADED APPROACH TO PREFERRED DISPOSAL TECHNOLOGY SELECTION, WITH INCREASING ISOLATION AND CONTAINMENT MOVING DOWN THE TABLE

Basic concept	Example design concepts	Waste groups	Example waste types	Remarks on waste properties
Near surface				
Earthen trenches	— Simple earthen trench with adequate natural or engineered barriers — Engineered cover	EW, VSLW	— Area building rubble, debris — Soil, large items and bulk waste	— Solid, chemically inert materials with very low radiation risk — Potentially use of existing landfills or suitably prepared construction pits/ quarries
	— Engineered earthen trench — Geomembrane and/or — compacted clay liner — Engineered cover — Leachate monitoring system — Design variants for wet and arid climates	VLLW, LLW	— Area building rubble — Soil, large items and bulk waste — Chemically impacted soils — Vehicles and large equipment — Biological waste (livestock, crops, trees, vegetative covers, etc.) — PPE, tools, rags — Incinerator ash	— Solid, potential for chemical contamination — Relatively short radioactive half-life and/or low radioactivity concentrations — Higher organic content (e.g. plant matter and putrescible)
Near surface engineered structures	— Concrete vaults, silos	LLW	— Reactor building rubble — Soil with potential for chemical contamination — PPE, tools, cleanup materials — Conditioned, solidified liquids and sludge — Incinerator ash	— Solid immobilized waste with potential for some contaminated bulk waste disposal — Predominantly short lived with limited long lived radionuclides — Higher radioactivity concentrations than in earthen trenches — Relatively low organic content and no free liquids
Geological disposal				

TABLE 14. GRADED APPROACH TO PREFERRED DISPOSAL TECHNOLOGY SELECTION, WITH INCREASING ISOLATION AND CONTAINMENT MOVING DOWN THE TABLE (cont.)

Basic concept	Example design concepts	Waste groups	Example waste types	Remarks on waste properties
Intermediate depth disposal	— Existing subsurface facilities (mines, tunnels) — Drifts, tunnels, silos — Borehole disposal concept	ILW	— Certain irradiated reactor internals and piping — Conditioned, solidified liquids and sludge — Reactor shielding rubble — Incinerator ash	— Long lived radionuclides and high concentration of short lived radionuclides — Relatively low organic content — Liquids if present in HICs
Deep geological disposal	— Purpose built deep geological repository	HLW, ILW	— Spent and damaged nuclear fuel — Irradiated reactor internals — Solidified high activity liquids and sludges	— Potential for heat generation — High concentrations of long lived radionuclides — Difficult waste with properties not conducive to long term containment

Note: PPE: personal protective equipment

Placement of a cap over the waste is typically used to limit water infiltration and control surface water drainage after disposal units are filled to capacity. Increased disposal capacity is often achieved by filling disposal units significantly above the original ground surface, using engineered earthen dikes and embankments to ensure adequate slope stability.

Co-disposal of multiple waste types may be acceptable, reducing the number and types of new disposal facilities needed. The USA, for example, allows slightly contaminated bulk soils, debris, demolition rubble and other VLLW to be co-disposed in authorized hazardous chemical waste disposal facilities.

11.4. SITING NEW DISPOSAL FACILITIES AFTER AN ACCIDENT

Experience suggests that siting new disposal facilities will present significant challenges. Depending on the situation, policy decisions might have to be made to develop new disposal sites on a compressed schedule. While certain siting tasks can be expedited, other tasks cannot be done quickly without compromising the safety case. Any expedited process would require the approval of the regulatory authorities. A balance is needed between an appropriate sense of urgency and allowing adequate time for all necessary work and public outreach to be undertaken properly. Timing objectives might require exceptions to be considered by regulatory authorities in order to expedite site characterization studies or other work, without compromising safety.

11.4.1. Siting disposal facilities in relation to the contaminated area

Whether sites ought to be sought inside or outside the contaminated area is a fundamental question that could arise. Making this choice can involve competing technical, economic and sociopolitical factors. Ultimately, decisions will be guided by specific accident circumstances and Member State policies. Questions to consider in evaluating alternative site locations include:

— Is the area near the accident site environmentally suitable for disposal based on permitting requirements and other regulatory standards?
— Would siting in the contaminated area compromise the ability to monitor new disposal facility performance and respond to unexpected releases adequately?
— Are any transportation or other cost savings that are achievable by disposing of waste near the accident site outweighed by asking the same communities experiencing the accident's greatest impacts to take on the burden of disposal as well? If so, might this be resolved through special economic benefits and incentives?
— If multiple facilities are needed, ought they to be co-located for economic efficiency or geographically dispersed so that the burden of hosting multiple facilities is shared more broadly? If so, what are the transportation impacts of dispersal?
— Are projected future land uses in the accident area compatible with new disposal facility development?

11.4.2. Local community incentives as a means to expedite new facility development

As noted above, offering benefits and incentives to help identify locally acceptable locations for new facilities has delivered mixed results. Experience is limited, however, so the approach deserves consideration. Offering incentives might expedite the siting process substantially if a volunteer community emerges and support for the facility continues through the time period needed to obtain necessary authorizations to construct a new facility.

Conversely, valuable time could be lost if no volunteers emerge, or if an initial community's willingness to host a new disposal site is later withdrawn after project mobilization. The impact of the latter scenario might be mitigated by identifying multiple volunteer hosts. A failed volunteer process could also create adverse perceptions that might increase opposition in other communities when the siting process is redirected and started anew.

Since offering incentives offers no guarantee of success, it might be advisable to proceed on multiple parallel tracks for siting a new disposal facility, including processes that do not rely on a volunteer local community.

11.4.3. Application of siting criteria in an accident situation

The normal process of implementing multiattribute site suitability criteria in order to identify preferred disposal sites typically includes stepwise application of exclusionary and discretionary components. Exclusionary criteria include characteristics that would eliminate broad siting areas or individual sites from consideration for waste disposal. Discretionary criteria are then used to distinguish more favourable from less favourable locations.

While technical judgements are necessary based on data availability, exclusionary criteria tend to be more absolute. In general, problems presented by exclusionary factors cannot be solved by reasonably available engineering measures. Exclusionary technical characteristics for near surface disposal typically include surface water bodies, floodplains, active or potentially active earthquake fault zones and areas of significant erosion or mass wasting, artesian or shallow groundwater conditions, oil or mineral resource extraction areas and projected population growth. Exclusionary environmental characteristics might include wetlands, designated wildlife or rare plant habitat and wild or scenic areas. Finally, exclusionary cultural and land use characteristics might include archaeological and historic sites, parks and recreation areas. In the case of a nuclear accident, it might be justified to site in areas that would normally be excluded in a non-accident context. For example, if there were to be a need for permanent relocation of affected residents (in a long term exclusion zone), this might allow consideration of a siting area that would be otherwise excluded due to the previously existing population. Conversely, projected population growth in the area could eliminate environmentally favourable sites from consideration.

Discretionary criteria are based on relative importance, as determined by sociopolitical values and practical limitations. Discretionary technical characteristics involve conditions that are amenable to engineered barriers or other technical solutions but are still useful for screening more favourable from less favourable sites. A variety of techniques have been used to weight or rank the importance of discretionary siting factors to help guide the screening of non-excluded siting areas to find favourable locations. Favourable technical conditions might include deep, confined, or non-potable groundwater; deep or low permeability surficial soils; non-porous bedrock; gentle topography. Examples of discretionary environmental, land use and cultural characteristics might include agricultural and forest lands, utility availability and local land use plans. Access by available roads and rail service can be an extremely important discretionary factor.

11.5. DISPOSAL FACILITY OPERATIONS FOLLOWING AN ACCIDENT

Accident wastes that are dissimilar to wastes generated for disposal under routine conditions could require additional WAC. New facility specific requirements might need to be approved by responsible regulatory authorities. Multiple agencies could be involved in addressing non-radiological waste constituents.

Demonstrating compliance at the time of waste acceptance at the disposal facility tends to rely on a combination of physical inspection and documentation prepared under an established quality management programme. Depending on the emergency waste types involved, physical inspection might include opening of packages, waste sampling and laboratory testing. Lessons learned from previous accidents suggest that proper implementation of WAC requires detailed procedures and rigorous, independent oversight to avoid improper disposal. Poorly controlled waste acceptance screening can significantly compromise disposal facility performance.

WAC are also applicable and important for waste storage facilities or temporary staging areas that hold and later ship waste for disposal, as discussed in Section 3.1. Failure to establish and adequately apply such criteria at these locations can risk extended delays in preparing wastes for shipment, rejection of shipments at the disposal site and other adverse impacts. For this reason, it is important that disposal facility operators ensure that waste storage and other predisposal operations are also governed by suitable waste screening and quality management programmes.

11.5.1. New disposal facility construction

Construction of a disposal facility for accident waste could differ from construction under non-accident circumstances in several respects. Existing experience demonstrates that the following practices might be undertaken to accelerate construction without compromising safety, if properly planned and carried out:

(a) Disposal units could be constructed in discrete, extendable phases, allowing an initial disposal area to be built and put into use, followed by later construction of new, connectable phases. Design of so called continuous lined disposal trenches is employed at the hazardous and VLLW disposal facility near Grand View, Idaho, USA [168].
(b) Tightly scheduled construction sequencing, beginning with advance mobilization of necessary construction equipment and personnel to the site, can help expedite schedules. Standard project management tools exist in the construction industry to accomplish this.
(c) Supporting utility, road, parking, fencing, equipment maintenance facilities and other support infrastructure can be constructed during the disposal unit permitting process.
(d) Other support facilities, such as mobile administrative and security offices, and worker change and shower facilities, can also be mobilized to the work site prior to construction.
(e) Coordination with regulators promotes timely review of post-construction as-built drawings, minimizing the time between construction completion and authorization of waste emplacement operations.

Special measures generally apply if construction takes place on contaminated land. This might include the following considerations:

— Implementation of a radiation dosimetry programme for construction workers, along with appropriate training;
— Procedures for managing contaminated clothing, tools and equipment — these might include a combination of decontamination and disposal of such materials.

11.5.2. Transportation to an off-site disposal facility: routes, modes and equipment

The large volumes of waste associated with an emergency may overwhelm a Member State's normal operating practices to load, transport, inspect and unload bulk and other wastes at the designated disposal site. Substantial experience exists concerning handling large volumes of waste, as described further below, and specific examples for accident wastes are presented in Section 9.

Choosing between road and rail or a combination of them using intermodal waste shipping containers is an important aspect of disposal planning. Road conditions, including the potential need for improvements and vehicle weight limits need to be appropriately considered. If rail is used, new sidings for exiting and re-entering the main rail line could be needed. Further, waste loading and unloading facilities will be needed at both the shipping and receiving ends.

Shipping waste by rail can accommodate large volumes on an expedited schedule, particularly if unit trains set aside solely for waste shipments are utilized. Solid wastes can be transported in gondola cars covered with tarps or hard lids. Waste can be wrapped within the railcar in a plastic liner, which facilitates waste loading at the accident site.

Shipping by truck can also accommodate large volumes. Bulk waste is typically covered by a tarp. Load capacity can be increased through the use of multiaxle haul trucks with a connected trailer, which distributes the total weight. Road weight limits could be increased temporarily in response to an accident, with appropriate government approvals. Depending on direct radiation levels in the waste, shielding might be needed in truck cabs to limit doses to drivers and a personnel dosimetry programme could apply.

A variety of shipping containers might be utilized, depending on the type of waste. For high radiation dose packages, certified transportation casks could be needed. Large reactor components might also be shipped intact by barge, rail or specialized trailers.

11.5.3. Shipment loading and unloading facilities

Compatible loading and unloading equipment and facilities will be needed at the accident site, the disposal facilities and any intermediate locations, such as temporary staging, interim storage, or processing facilities. Depending on the waste volumes handled and the transportation modes utilized, this may require large areas. On-site rail facilities, for example, may beneficially include multiple track extensions to hold railcars pending unloading, cleaning, inspection and release back to the main regional rail line.

A variety of equipment is readily available to transport and unload waste. Bulk waste transported by truck can be hauled by end dump, side dump, flatbed trucks hauling intermodal or other boxes, and covered trailers containing drums or other containerized wastes. For high dose rate packages, unloading by crane or other remote handling equipment might be appropriate.

Unloading facilities are governed by the training, industrial safety, QA, personnel dosimetry and other radiation protection programmes and procedures that apply to the rest of the disposal facility.

11.5.4. Verification of WAC at the disposal facility

Experience indicates that receiving large numbers of accident waste shipments can strain a facility's ability to inspect all incoming wastes adequately. Adapting to these circumstances generally requires expanded personnel and training, close supervision and increased inspection frequency to ensure that procedures are followed. Increased QA programme audits are another means to identify areas warranting improvement. Activities meriting special attention might include the following:

- Waste verification coordination between personnel at the disposal facility and all predisposal management facilities shipping waste;
- Review of manifest shipping documents to verify that WAC are met;
- Direct radiation measurements of incoming shipments, as well as equipment leaving radiation controlled areas;
- Visual inspections and waste sampling and testing to verify waste characteristics;
- Increased instrument calibration frequency;
- Documentation and recordkeeping for wastes received and location disposed at (a disposal unit grid system is commonly used) to assist retrieval, if needed.

To minimize rejection of waste shipments and further ensure compliance with WAC, placing trained personnel at the accident site and any intermediate waste transfer or processing facilities can be beneficial to oversee shipment preparation and perform independent oversight. Preapproval of the QA programmes used in such off-site locations could also be appropriate. Through these measures, the disposal facility operator can better determine the following:

(a) What the source of the waste was in the accident zone;

(b) Whether the waste was adequately characterized and segregated from other, incompatible waste types;
(c) Whether possible processing treatment was properly carried out;
(d) Whether direct radiation levels on bulk shipments and waste packages comply with the limits established for the disposal facility.

11.5.5. Operational monitoring and surveillance at the disposal facility

Depending on the waste streams resulting from an emergency and the disposal facility design, monitoring and surveillance programmes might need to be upgraded. Areas potentially requiring such attention could include the following:

— Expanded groundwater monitoring to account for hazardous chemical or other non-radiological constituents of concern;
— Gas generation resulting from degradation of organic wastes;
— Fire protection considerations, if combustible waste volumes increase or significant methane gas is generated over time;
— Industrial safety considerations related to potential use of new waste handling equipment, exposure to additional hazards, such as chemical substances or infectious wastes, and other worker safety concerns.

11.6. EXPERIENCE WITH DISPOSAL FOLLOWING MAJOR ACCIDENTS

Aside from the major accidents used as examples in this publication, the most extensive relevant experience of disposal of large volume wastes arising in contaminated areas comes from the management of legacy wastes from historic nuclear weapons development activities. Examples include the remediation of US nuclear weapons related facilities, Sellafield in the UK and the Maralinga, Australia cleanup (see Appendix VI).

When waste volumes are relatively small and disposal facilities exist to accept the waste, disposal outside of the contaminated area has often been a preferred option. Examples include the disposal of remediation waste from smaller US nuclear weapons facilities (e.g. near surface repository and bore hole disposal at the Nevada National Security Site), and the WIPP project, and waste from contaminated facilities in urban areas, such as the Goiânia, Brazil abandoned source dispersal incident (see Appendix V) [169].

A combination of disposal within and outside contaminated areas has also been implemented. This has typically involved the excavation and removal of more highly contaminated wastes for off-site disposal, allowing more lightly contaminated materials to be disposed of near the site of the emergency. Examples of this strategy include the Maralinga, Australia and Palomares, Spain disposal programmes [75, 170–172].

The emergency response to the Chernobyl accident led to rapid waste disposal without prior waste characterization or detailed disposal facility engineering [3]. Follow on remedial actions are likely to continue for decades. Widespread experience also exists with remediation of legacy disposal in non-engineered disposal settings without safety case preparation. Preplanning can provide a framework for evaluating site characteristics and identifying minimum engineering design standards needed in the event of an accident.

Experience with legacy practices indicates that failure to establish or consistently apply WAC can result in the need for extensive future remediation. This has involved waste recovery by exhumation and subsequent redisposal, a hugely expensive and time consuming process with significant risk of increased worker exposure to radiological and other hazards. Waste removal and reburial has occurred extensively in the remediation of weapons related sites in the USA.

Experience with the timing of disposal varies widely. IAEA Safety Standards Series No. SF-1, Fundamental Safety Principles [173] discourage leaving future generations with the burden of radioactive waste disposal, implying timely, safe disposal of wastes. The ability to avoid passing on burdens after an accident is a challenge in the case of both the Chernobyl and Fukushima Daiichi accidents, where waste disposal is taking place over extended time periods due to the large volumes and diverse waste streams, and new disposal capacity needs to be developed. Delayed decommissioning has also been utilized following the TMI-2 and Windscale Pile I emergencies. In the case of small scale nuclear accidents and remediation of legacy sites, rapid disposal has occurred using existing disposal capacity.

Once plans are made to proceed with disposal, a different set of timing challenges arise. If existing disposal capacity is adequate to receive all accident related wastes, tasks and schedules can be set with reasonable confidence, provided that sufficient financial and human resources are available. Here, extensive experience is available, applying generally accepted project management tools to planning and implementing the sequencing of siting as well as waste characterization, treatment, processing, loading, transportation, receipt and disposal of large waste volumes.

In cases where disposal capacity and sufficient funds are unavailable, the national decision has been to delay disposal. In Japan, a decision was reached following the Fukushima Daiichi accident to rely on long term storage pending future identification of preferred disposal locations and technologies. In Ukraine, lack of financial resources has limited disposal progress.

12. CONCLUSION: THE VALUE OF BEING PREPARED

It is impossible to predict the nature and scale of any future nuclear accident. However, there is now extensive experience in some Member States concerning the impacts of major accidents and the effectiveness of responses to them. The practical experience and lessons learned regarding managing accident wastes that are presented in this publication provide other Member States with a good basis for preplanning of options and possible solutions for accident waste management. That experience can be combined with decades of knowledge gained from managing large volume wastes from cleanup of major contaminated legacy sites and from NPP decommissioning. Together, this information provides a sound basis for understanding the waste management issues that will be involved in responding to an accident.

A key lesson learned is that safe, cost effective management of waste from a major nuclear accident has proven to be a daunting task that could have been simplified had there been advance contingency plans in place at both national and facility levels. Preplanning and preparations can be made by Member States that will significantly improve the effectiveness of waste management and site restoration in the event of an accident. Such preplanning can be based on the identification and analysis of potential accident scenarios for specific nuclear facilities and could involve regional and local waste management options studies and exercises involving the organizations that would be involved in accident response. Preplanning can guide response actions to ensure that sound technological and programmatic solutions are identified in the aftermath of an accident and can direct contingency actions as the events of an accident unfold and the on- and off-site conditions change. These preplanning efforts will also support stakeholder and public understanding of the options and approaches that will be available for waste management.

A clearly defined waste management strategy and planning process is thus imperative for success. Without a clear strategy, response actions taken in the early aftermath of an accident can limit the range of future management options, substantially increase costs and result in significant worker exposures. A preplanned waste management strategy will provide a sound framework for the decision making process for technology selection and waste management and disposal facility siting, which will need to function under extreme time pressures at the inception of an accident, as well as over the long duration of site cleanup and remediation.

Through robust preplanning, Member States can also prepare themselves to minimize the amount of waste requiring disposal, separate wastes by type and radioactivity level, and process or otherwise prepare stored waste for disposal and then dispose of it in a safe, efficient, cost effective manner. It is paramount for the waste management programme that all work be conducted in a manner that protects workers, the general public and the environment, in accordance with accepted standards.

Appendix I

THE WINDSCALE PILES ACCIDENT

The fire at the Windscale Pile 1 reactor in 1957 was well documented, both at the time and in subsequent studies. This appendix provides a brief outline of the accident and the management of the resulting wastes.

I.1. THE WINDSCALE PILES REACTORS

The Windscale Works (now part of the Sellafield site), located on the west coast of Cumbria in the UK, were established in 1947 with two 180 MW(Th) graphite moderated, air cooled reactors being constructed for the production of plutonium for national defence purposes and radioisotopes for medical, military and industrial applications. The natural uranium fuel rods were clad in finned aluminium cans. Pile 1 went operational in October 1950 (Fig. 55).

Routine annealing, by raising the reactor core temperature, was carried out periodically between 1952 and October 1957 to remove stored Wigner energy in the graphite. The accident at Pile 1 occurred during one of these annealing exercises.

FIG. 55. Windscale Piles 1 and 2. Courtesy of Radioactive Waste Management Ltd.

I.2. THE PILE 1 FIRE

On 10 October 1957, an annealing exercise resulted in an unexpected rise in core temperature. An attempt to cool the reactor core by blowing air through the fuel channels led to some of the channels catching fire. Fuel was removed manually from the burning channels and an unsuccessful attempt made to extinguish the fire with CO_2. Eventually, the fire was put out by pumping ~7000 m^3 of water through the core. This water was discharged into the sea or the surface water drainage system. The accident is described in detail in Refs [60, 174, 175], with a recent evaluation of the consequences and impacts being presented in Ref. [176]. The report of the Committee of Inquiry appointed immediately by the UK Atomic Energy Authority was produced two weeks after the accident. Led by Sir William Penney, the committee report (the 'Penney Report') was subsequently published in 1988 and is now available in Ref. [177].

The fire caused the airborne release of radioactivity from the pile stack. Although this was fitted with filters that proved highly effective, a plume of contamination spread off-site. Throughout the period of release, measurements of air contamination levels were made both on the site and in the district. These measurements were continued throughout the period of release and it was apparent that the accident had not caused any significant external radiation exposure to individuals living in the district and that the external exposure incurred by the public living in the neighbourhood was negligible [178]. At a greater distance, two plumes of activity of similar magnitude extended to the south and the east, with significant deposition of ^{131}iodine across the English Midlands and contamination being detected as far away as Belgium.

As a result of the fire, the hazard to the public arose from inhalation, ingestion and external radiation arising from fission products, with the principal health hazard to workers and the public being from radioactive iodine. Thyroid surveys were performed in order to assess the effects and the highest activity was found in the thyroid of a UKAEA employee, at 0.5 µCi. Surveys were also carried out to assess worker exposures to ^{89}Sr/^{90}Sr. The highest levels detected were found to be no more than one tenth of the maximum permissible body burden. Radioactive caesium levels were found to be acceptably low in biological samples [179]. It was possible at an early stage to reject the need for emergency measures based on inhalation or external radiation. Possible ingestion hazard arising from contaminated pastureland was tested by collecting milk from local farms. The ^{131}iodine content made it necessary to restrict the distribution of milk from around the site, eventually in an area of ~500 km^2. Approximately 3000 m^3 of milk was collected and disposed[4]. Drinking water supplies in Cumberland, Lancashire and North Wales were also analysed for activity and it was found that the level of contamination was well below what would constitute a hazard [178].

I.3. MANAGEMENT OF SOLID WASTES

Both reactors remained shut down without any further annealing of their cores and the Pile 1 reactor was deemed unsalvageable. The initial phases of cleanup commenced in 1958 and were largely completed by 1959 [180]. Much of the fuel was removed from Pile 1 and sent for storage and/or reprocessing.

In the initial cleanup, solid wastes, which included the stack filters and various debris from Pile 1, were sent to available facilities on the Windscale Piles site, including the Solid Waste Storage Silo (used for the storage of fuel decanning materials), the Windscale Disposal Trenches for LLW burial and the North Group Waste Store (a brick walled flat roofed structure where waste was interred). There are few records of the disposals to these facilities and current inventories have to be estimated from other historic records, including those held at the National Archives and from consigning plants. In 1981 it was decided that there was no risk of core fire and no criticality problem from core collapse, allowing decommissioning to commence in the mid 1980s. Phase 1 decommissioning was completed in July 1999.

[4] See: Nuclear Power and the Environment, https://pubs.rsc.org/en/content/ebook/978-1-84973-194-2.

The final decommissioning of the two piles is not completed. A range of proposals and some early decommissioning activities, combined with several changes in organizational responsibilities for the site and continued enhancements in decommissioning management techniques, have led to several changes of plan. Latterly, this has involved increased stakeholder involvement, in particular concerned with deciding the desired end state of the site. A key issue is the materials characterization required prior to detailed design development and the start of physical works that would generate wastes. Characterization supports waste management planning to identify waste sentencing routes and the R&D and safety cases that are required to perform the decommissioning work.

The decommissioning scheme for Pile 1 involves a structure built over the pile cap with a slot cut in the cap to facilitate the deployment of tools and equipment. This will initially allow retrieval of the fuel and isotope cartridges currently remaining in the core and their subsequent processing. The cartridges are placed in mild steel liners and subsequently encapsulated within the liner using an organic polymer. The liner will then be grouted into a 500 L stainless steel drum, creating a grout annulus between the drum and the encapsulated waste. The 500 L drums will be placed into interim storage in a Sellafield shielded store to await eventual disposal in the planned national GDF. Polymer was selected over conventional cementitious grouting because, under alkaline conditions, metallic uranium can react to produce voluminous and reactive corrosion products.

It is anticipated that a number of patterns of liner will be required to accommodate the different nature of various components of the fuel and isotope waste stream, such as a higher degree of shielding for cobalt cartridges, and it is intended to run operations in campaigns, with each campaign involving the removal of cartridges of a particular type, along with the packaging of items already retrieved and in storage. The condition of fuel and isotope cartridges remaining in Pile 1 after the fire is variable, ranging from intact to ash and debris. The ash is considered to be a combination of oxides and hydroxides of uranium, cladding and isotope source materials.

The Pile 1 graphite will be put into in baskets and placed inside 3 m^3 boxes (Fig. 56), which will also be exported to a Sellafield shielded store to await eventual disposal in the national GDF. Until issues regarding Wigner energy and the potential need to anneal the graphite are resolved, the retrieved graphite will not be encapsulated. To keep the options of encapsulation and annealing open, the waste baskets will be designed so that they can be retrieved from the 3 m^3 boxes if this is shown to be necessary, and packed with waste in a manner that will allow grout infiltration at a future time, while eliminating unnecessary voidage.

Until the national GDF is available, the ILW from decommissioning Pile 1 will be placed in interim stores on Sellafield site. This is consistent with the preferred options of previous option selection studies and complies with the waste management proximity principle. It allows waste containers to be placed directly into a safe storage location and eliminates the need to prepare ILW packages for transport on public roads at this time. The Nuclear Installations Inspectorate has recommended that radioactive waste and material being placed in storage now will require containment for an overall period of at least 150 years [181]. In order to maintain the required longevity of the 3 m^3 boxes and 500 L drums, the atmosphere of the intermediate store will be controlled, to limit chloride and control temperature and humidity.

FIG. 56. The 3 m³ box for Pile 1 graphite and other reactor wastes that will eventually be disposed of in the national GDF. Courtesy of the Nuclear Decommissioning Authority.

Appendix II

THE THREE MILE ISLAND ACCIDENT

TMI NPP is located on Three Mile Island on the Susquehanna River in Londonderry Township, Pennsylvania, USA. The plant has two separate pressurized light water reactor units, TMI-1 and TMI-2, and came on-line in 1974. TMI-1 continued in operation until it was permanently shut down in September 2019. The accident occurred on 28 March 1979 in Unit 2 (TMI-2). This appendix gives a brief description of the accident and the management of the wastes that were produced.

II.1. THE TMI-2 ACCIDENT

The accident was caused by several mechanical failures that were compounded by plant operator failure to recognize the loss of coolant accident conditions and uncertainty in the status of system valves. The result was a partial meltdown of the reactor core. Fortunately, very little radioactive material was released from containment and the public health and environmental impacts were minimal. However, cleanup after the accident took over 14 years, at a cost of over US $1 billion.

Approximately 2 million people lived around TMI-2 during the accident and are estimated to have received an average radiation dose of only ~0.01 mSv above the usual background dose [42]. Various government agencies assigned to monitoring the area collected environmental samples of air, water, milk, vegetation, soil and foodstuffs. Very low levels of radionuclides were found to be attributable to releases from the accident. However, in spite of serious damage to the reactor, the actual release had negligible effects on the physical health of individuals or the environment (Table 15).

Today, the TMI-2 reactor is permanently shut down and 99% of its fuel has been removed. The reactor coolant system has been drained and the radioactive water decontaminated and evaporated. Radioactive waste from the accident was shipped off-site for disposal or storage. After waste had been removed, TMI-2 was placed in long term, monitored storage pending the permanent shutdown of TMI-1 [42]. In December 2020 the licence for TMI-2 was transferred to a decommissioning contractor [56].

TABLE 15. SOLID AND LIQUID WASTES RESULTING FROM THE THREE MILE ISLAND ACCIDENT

THREE MILE ISLAND	
Solid waste (type and quantity)	Liquid waste (type and quantity)
DAW (concrete, core debris, decontamination materials, job control waste, etc.): ~3900 m³ total (36% compacted, 60% non-compacted, 3% stable, 1% other) Wet solid wastes (ion exchange materials, sludges, etc.): ~1200 m³ total Class A: 93%; class B: 4%; class C: ~3%; greater than class C: ~1%	Low activity water (<1 µCi/mL): non-accident water Intermediate activity water (1–100 µCi/mL): fuel handling building floor and reactor coolant bleed tanks High activity liquid wastes (>100 µCi/mL): reactor building basement and makeup and purification systems Total 7500 m³

Note: In US waste classifications, classification is dependent on the concentrations of certain radionuclides. Class A: background to 700 Ci/m³; class B: 0.04 to 700 Ci/m³; class C: 44 to 7000 Ci/m³.

II.2. MANAGEMENT OF THE ACCIDENT WASTES

The EPRI documented the challenges and decisions made in the aftermath of the accident and reviewed the questions and issues that were identified, the options evaluated for cleanup and waste management, the actions taken and the consequences of those actions [182]. A further EPRI report provides detailed technical and operational information pertaining to all aspects of waste management after the accident [183]. The information contained in these and other reports is summarized in this appendix.

Solid, liquid (primarily contaminated water) and molten fuel core wastes were generated after the accident. Over 5000 m^3 of solid radioactive wastes (dry and wet) were managed over the course of the cleanup. Over four million litres of contaminated water were treated using ion exchangers and sorbent systems to remove radionuclides, resulting in wet solid wastes requiring disposal. Approximately 140 000 kg of core debris was retrieved from the reactor and transported in 52 shipments to the Idaho National Laboratory, where it remains today in dry storage.

Certain wastes exceeded US commercial burial limits and could not be disposed of in existing sites. A Programmatic Environmental Impact Statement evaluation was conducted and an agreement was made between the US NRC and the US DOE for the DOE to accept these wastes for R&D purposes, to evaluate treatment technologies, storage systems, etc.

Cleanup progressed in four overlapping operational phases from 1979 to 1990: stabilizing the plant, waste management, decontamination and defuelling. Safety was the overriding guide in planning and conducting activities. A decision not to restart the reactor was made in the early stages, which helped to focus waste treatment and decontamination activities. A strategy to pursue flexible and parallel cleanup technologies was employed to avoid project stoppages and minimize project restarts. The management team also worked with regulators and stakeholders to develop acceptable solutions.

Acquisition and analysis of materials characterization data were deemed vital to support informed decision making on plant stabilization, cleanup approaches and waste management. Extensive on-site analytical capabilities were established, although numerous samples also had to be shipped off-site to other facilities for analyses. The transport of these materials became a central issue and a packaging coordinator function became an important element of the cleanup process.

There was essentially no external radioactive material release during or following the accident, so solid waste management primarily involved handling of decontamination equipment and media, water treatment ion exchangers and sorbents, and debris and fuel materials associated with fuel in the reactor. The primary waste management concern following the accident was the retrieval, treatment and handling of contaminated water, for which water treatment systems had to be developed. The EPICOR II system was constructed in the seismically qualified Chemical Cleaning Building and was designed to facilitate easier insertion and removal of liners containing ion exchange media using a monorail. It removed approximately 3×10^{15} Bq (80 000 Ci) of activity from ~4 000 000 L of water, using organic, zeolite and charcoal media. Various water management systems also had to be developed to treat high level liquid wastes (contained in the reactor building basement and the reactor coolant and purification systems that had activities >100 μCi/mL water) and to remove solids and maintain Cs and Sr levels within acceptable limits during defuelling, using zeolite media and filtration systems to remove particulate matter. Providing sufficient water storage capacity was a significant issue during cleanup.

Very little solid waste storage capacity existed prior to the accident. However, once cleanup commenced, solid waste began to accumulate quickly on the site. The early solid wastes were primarily trash and spent ion exchange materials. Later, spent ion exchange system prefilters and resin filled liners dominated solid waste activity levels. Regulatory and political issues in the aftermath of the accident made shipping the waste off-site for final disposal problematic and temporary storage facilities became necessary. These were designed to be flexible, to accommodate waste storage and staging for off-site shipping. Immediately after the accident, existing facilities were reconfigured for waste storage, but the need for an engineered temporary storage facility was soon apparent. In 1980, a storage facility known as 'waste acres' was put into service. This modular facility consisted of two storage modules with a common drainage sump. Each module contained 6 rows of 10 cells (Fig. 57). The module walls were 1 m thick with a 1 m thick lid to provide shielding. Each cell had a 2 m diameter and was 4 m high, with a drain line into the sump. A crane system was used to move ion exchange liners and storage drums in and out of the cells.

The Interim Solid Waste Staging Facility was constructed in 1982 to provide an additional storage area for LLW drums (208 L) and low activity waste boxes. Again, the design allowed for ready shipment of materials to off-site disposal facilities. A Waste Handling and Packaging Facility became operational in 1987, to consolidate and package solid wastes and for decontamination. The facility significantly reduced waste volumes (25–30% improvements in packing efficiency) and allowed commercial release of decontaminated materials. Compactible waste was volume reduced using a stock drum compactor.

Regulatory and political issues precluded disposal of wastes immediately after the accident. Initially, the governors of two states rejected the disposal of TMI-2 wastes in the LLW disposal facilities at Hanford (Washington state) and Barnwell (South Carolina). The governor of Washington allowed wastes to be shipped to Hanford in late 1979, but it was not until 1987 that LLW was also accepted at the Barnwell facility.

Higher level wastes provided a challenge, but an agreement was made between the US NRC and the US DOE for the latter to accept the 'greater than class C' wastes for R&D purposes to evaluate treatment technologies, storage systems, etc. This involved several types of wastes, including fuels, EPICOR II resins, filters, TRU contaminated materials and liners used in specialized decontamination stations (SDSs). The US NRC waived its requirement to solidify the EPICOR II resins in 1981 and efforts were made to develop means to package the resin liners for disposal at DOE sites. The resulting concrete reinforced HIC (see Fig. 58) proved to be acceptable for disposal of the EPICOR II resins at the Hanford site.

The HIC contained a coated, corrosion resistant steel liner. Buffering material to control pH was added to the bottom and the container was closed by bonding the lid to the body using an adhesive gel and flowable grout. A vent system allowed radiolytic gases to escape. Extensive evaluation was also carried out concerning ways to dispose of the wastes from SDSs; the systems used to demineralize the high activity liquids in the reactor building basement. The primary waste from the SDSs consisted of liners. For disposal purposes, the SDS liners were packaged in polyethylene lined HICs and transported to the Barnwell site for disposal, once shipments to the Barnwell site were allowed.

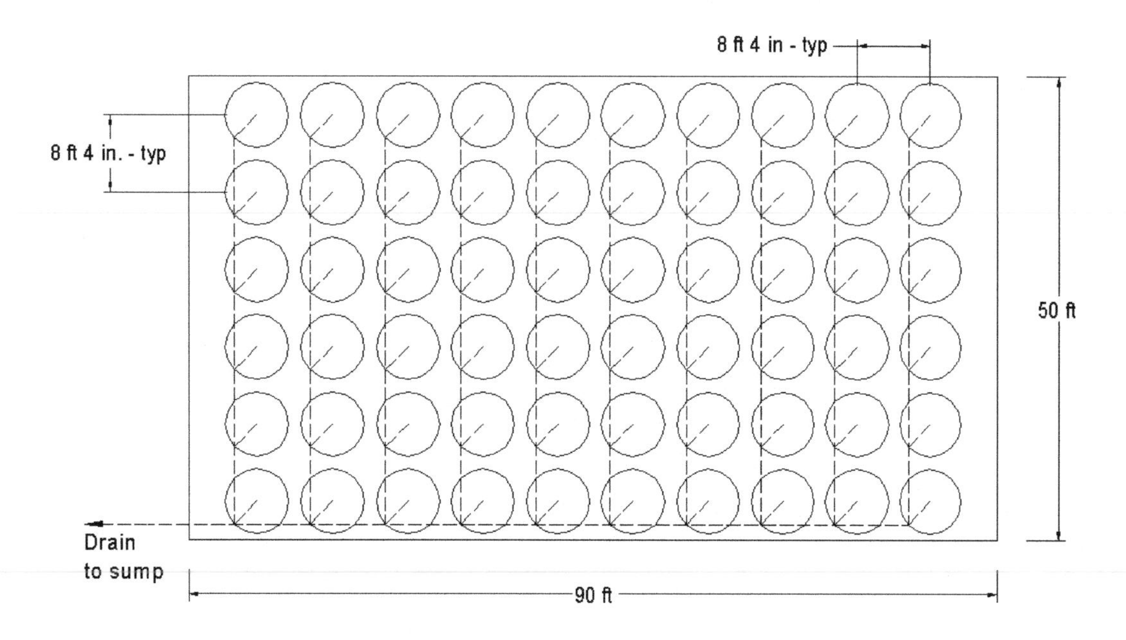

FIG. 57. The solid waste staging area constructed at TMI-2. Adapted from Ref. [2].

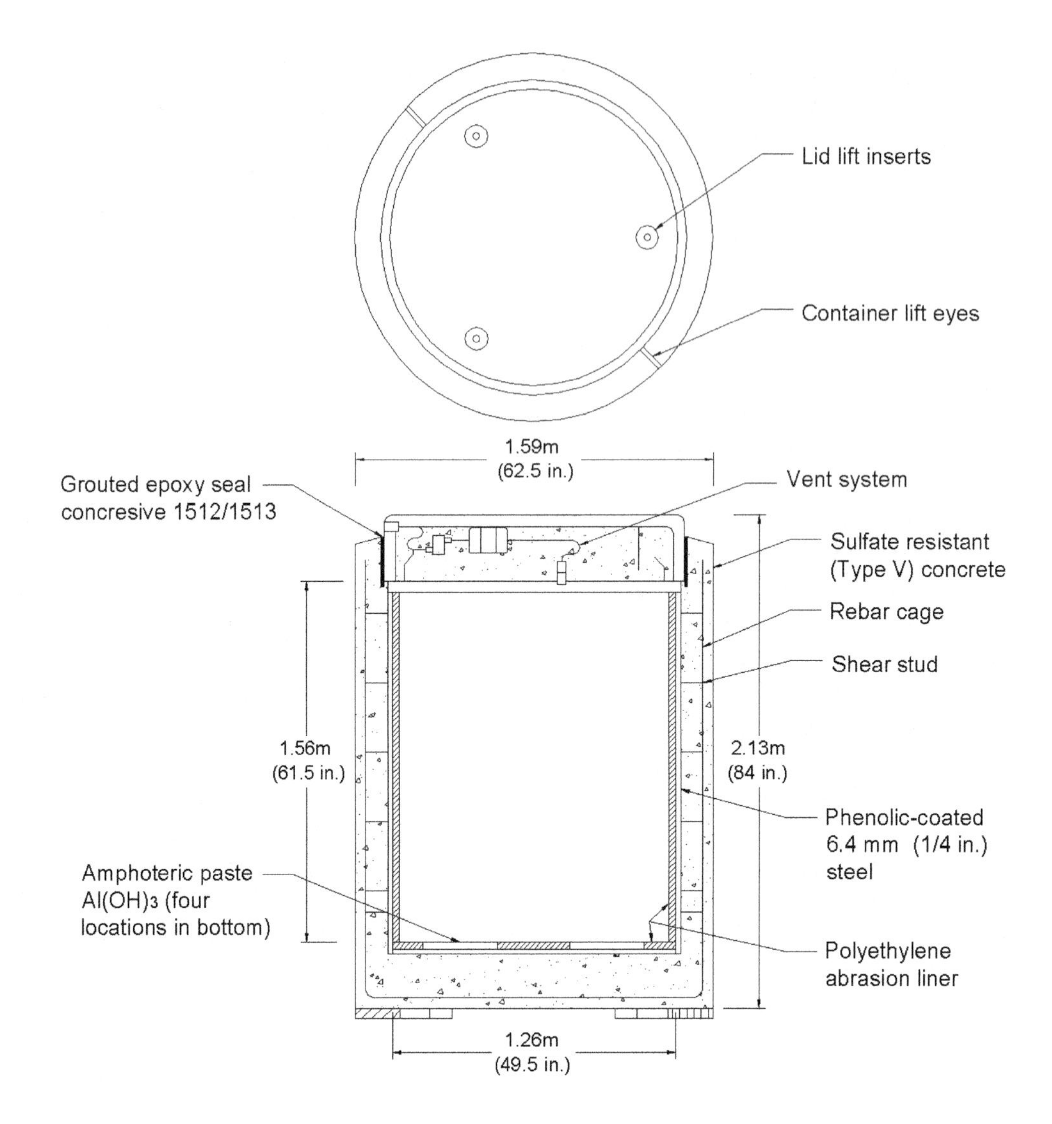

FIG. 58. The reinforced concrete high integrity container design. Adapted from Ref. [2].

Appendix III

THE CHERNOBYL ACCIDENT

In April 1986 an explosion and fire destroyed the Chornobyl NPP near Kiev, in the former USSR. It is acknowledged to be the worst accident ever to have occurred at a nuclear power facility. It caused immediate and early deaths and led to severe contamination across wide areas of what are now Ukraine and Belarus, with low levels of contamination detectable across much of Europe. A large local population had to be permanently moved and an 'exclusion zone' established in the most contaminated areas, which is expected to exist into some indefinite future time. The impact of the accident on the local communities and on the development of nuclear power, both in the former USSR and worldwide, has been immense. Remediation at the site continues more than 30 years later and will extend many more decades into the future.

This appendix briefly outlines the considerable work that has been undertaken to manage the wastes generated on- and off-site, from immediately after the accident to the present. Both the waste management programme and the accident itself, along with its health consequences, have been recorded and discussed in numerous studies and publications, the depth and detail of which are beyond the scope of this short appendix to capture in full.

III.1. THE CHERNOBYL ACCIDENT

Reactor No. 4, where the accident occurred, was one of four high-power channel-type reactors (RBMK), light water cooled, graphite moderated reactors at the Chornobyl NPP. The accident was caused by a combination of poor reactor safety system design and operator failure [184]. It occurred during a test intended to improve reactor safety in the event of a loss of power. Several of the critical safety features of the reactor had been disabled in order to perform the test, which went out of control so rapidly that operators could not prevent a massive explosion. One model suggests that an initial explosion in channels beneath the reactor tore the 1000 t lid off the containment and lifted the core 30 m, completely out of the reactor shaft. The core exploded in the air within the reactor hall, destroying the building and throwing fragments of fuel into the immediate environment of the reactor buildings and the wider NPP site, with one fuel assembly reportedly having been found 1600 m away [185]. Approximately half of the core fell back into the reactor shaft. During the explosion, part of the fuel and other components melted and were forced into parts of the reactor building, solidifying as 'lava' (or 'corium'). The exposed graphite core caught fire, causing the bulk of the airborne contamination that spread across Europe and the northern hemisphere. Initial attempts to extinguish the fire by dumping thousands of tonnes of loose solids (sand, lead, crushed dolomite, etc.) into the reactor from helicopters were unsuccessful and added to the spread of airborne particulate contamination. The graphite burned out and the fire was extinguished nine days later.

In the seven months after the accident, a shelter (the Shelter Object or 'sarcophagus') was built to protect the exposed remains of the reactor building from the elements and allow some remediation work to begin. The reactor and the Shelter Object (which was in an increasingly unstable condition) have now been completely covered by the NSC system: an arched structure that was completed with international funding in 2018. Reactor No. 4 and its Shelter Object will be dismantled within the NSC.

III.2. THE ACCIDENT WASTES

In the three decades since the accident, cleanup activities have generated a huge amount of waste, which is currently managed in 17 storage or disposal facilities located on the site or elsewhere in the exclusion zone (see Table 16). Together, they hold over 13 million cubic metres of material. The additional LLW that will be generated by demolition of Reactor 4 and the old Shelter Object might be disposed of on-site, with long term storage inside the NSC for ILW and HLW. The future of the fuel wastes is yet to be decided and there is still not a completely agreed picture of where all of the fuel is located in and around the site.

As can be seen in Table 16, the bulk of the accident waste is contaminated soils within the exclusion zone. A particular problem is that some of these soils contain fine particles of fuel — a plume of fine fuel debris was deposited unevenly across the exclusion zone, predominantly to the west and north and in the vicinity of the NPP, but extending up to 100 km away in a narrow strip to the west [187]. The predominant, most widely spread contaminants in the exclusion zone are, however, fission product radionuclides. A more detailed breakdown of waste materials located in various facilities at the site and in the exclusion zone is given in Table 17.

Initial collection of wastes took place in the 18 months immediately after the accident as Shelter Object construction began and during decontamination of the whole industrial site, as preparations were made ready to restart the other three reactors at the NPP. The main period of decontamination in the wide region of the Exclusion Zone, including the evacuated, populated districts, began in late 1986 and ran until late 1988. During this period, much of the waste that had been placed in temporary storage was moved to long term storage sites.

The first stage of waste collection and disposal was the most difficult stage of the decontamination activities on the site, as it involved collection of core and reactor debris with extremely high levels of radioactivity. Military personnel, using armoured vehicles with added radiation protection, assisted in the collection of scattered debris into metal containers (initially, containers for household litter with lids) and the topsoil layer, together with a part of debris, was moved to the boundaries of the collapsed reactor building by bulldozing. The cleaned area was covered by precast reinforced concrete slabs. Shift times were short and the vehicles were decontaminated after each shift. Some metal containers with HLW were initially buried at a regular disposal site and covered with concrete slabs and soil, then a system was developed where the waste containers were placed behind so-called 'pioneer' walls around the edge of the ruined plant building and covered with gravel, sand and concrete. These eventually became part of, or were contained within, the Shelter Object.

TABLE 16. ESTIMATED TOTAL AMOUNT OF RADIOACTIVE WASTE IN THE FACILITIES WITHIN THE CHORNOBYL EXCLUSION ZONE (ADAPTED FROM REF. [186])

Type of storage facility	Number of facilities	Waste volume (m^3)	Activity (TBq)
Disposal sites	3	631 519	5 420
Temporary storage facilities	9	1 296 588	1 840
Shelter object (including industrial site)	1	662 500	740 000
NPP storage facilities	4	19 794	385
Exclusion zone soils	—	11 000 000	8 130
Totals	17	13 610 401	755 775

Temporary storage was developed to hold bulldozed and collected topsoil from the wider site area that had not been covered in concrete slabs. This HLW also contained fuel and reactor debris and was collected in 1-m^3 containers. Some was placed into an unfinished LLW disposal facility on-site [188]. Altogether, more than half a million square metres of the site was cleared of topsoil.

A particularly hazardous problem was the removal of many tonnes of reactor core material (including 16.5 tonnes of fuel assemblies and 182 tonnes of graphite) that had been ejected onto the roofs of the adjacent buildings, including the Unit 3 reactor — an area of about 3000 m^2. This was carried out manually, with material either being pushed into the destroyed reactor building void or collected in containers and removed by crane or helicopter to one of the storage sites. Some of the material was placed in bags around the perimeter of the buildings and concreted into place [188].

The adjacent Unit 3 facilities also required major internal decontamination in order to allow operation of the NPP, and Units 1 and 2 had also been severely contaminated by airborne radionuclides entering through the ventilation systems. More than 24 million m^2 of inner facilities and more than 6 million m^2 of the surrounding area were decontaminated during 1987–1988, resulting in 38 000 tonnes of contaminated equipment and over 16 000 tonnes of radioactive waste that required transport and disposal.

After the initial cleanup of the site, attention turned to cleaning up the areas within a 30 km radius that had the highest levels of surface deposition of radioactivity. The most highly contaminated areas were within 5 km of the NPP and included the evacuated town of Pripyat. Many temporary or 'soft' buildings with high levels of contamination on their roofs were demolished. It was not possible to decontaminate the large area of the town successfully. Decontaminated sections rapidly became re-contaminated [189].

TABLE 17. DETAILED BREAKDOWN OF WASTES RESULTING FROM THE CHERNOBYL ACCIDENT AND THEIR LOCATION [6]

RADIOACTIVE WASTE ASSOCIATED WITH THE SHELTER OBJECT			
Fuel-containing mass with a nuclear fuel content greater than 1%			
Radioactive Materials	RW type by dose rate or activity	Quantity	Mass (tonnes)
Fuel assemblies with fresh fuel	HLW	48 pieces	5.5
Fuel assemblies with spent fuel		129 pieces	14.8
Dust with a nuclear fuel content of > 1% (mass.) and particle size up to several hundred microns		---	≈50
Lava-fuel containing masses		---	1 250
Core fragments		No data	
Secondary uranium minerals		---	0.01
Solid radioactive waste in the Shelter Object, with a nuclear fuel content less than 1%			
Radioactive Materials	RW type by dose rate or activity	Quantity	Mass (tonnes)
Equipment, metal (dose rate >10 mSv/h)	HLW		23 240

[6] Data provided courtesy of Special State Enterprise Chornobyl Nuclear Power Plant.

TABLE 17. DETAILED BREAKDOWN OF WASTES RESULTING FROM THE CHERNOBYL ACCIDENT AND THEIR LOCATION[5] (cont.)

RADIOACTIVE WASTE ASSOCIATED WITH THE SHELTER OBJECT			
Fuel-containing mass with a nuclear fuel content greater than 1%			
Radioactive Materials	RW type by dose rate or activity	Quantity	Mass (tonnes)
Building structures, including concrete, reinforced concrete, slabs, beams, rubble (dose rate >10 mSv/h)		38 000 m^3	---
Graphite		---	≈700
Dust <10 mSv/h	ILW, LLW	--- ---	
Building structures, including concrete, reinforced concrete, slabs, beams, rubble (dose rate <10 mSv/h)		299 000 m^3	---
Equipment, metal (dose rate >10 mSv/h)		---	18 200
Non-metallic materials, including plastics, cables, thermal isolation, among others (dose rate >10 mSv/h)	---	5 000 m^3	18 200
Liquid radioactive waste in the Shelter Object			
Radioactive Materials	RW type by dose rate or activity	QuantityVolume	Mass (t)
Water solutions, pulp, oil and its suspensions (specific activity < 3.7×108 Bq/m^3)	ILW, LLW	2 500–5 000 m^3	
Water solutions with the content of uranium salts, pulp, oil and its suspensions (specific activity < 3.7×108 Bq/m^3)	ILW, LLW	500–600 m^3	
Solid radioactive waste located on-site at the Shelter Object			
Radioactive materials	RW type by dose rate or activity	Volume (m^3)	Mass (t)
Nuclear fuel buried in the ground	HLW		≈ 0.6
Containers with core fragments and other materials buried in the ground (along row A, axis 68)	HLW ILW	600 1 100	
Bulk soil (crushed stone, sand, gravel)	HLW ILW LLW	600 2 000 137 000	

[5] Data provided courtesy of Special State Enterprise Chornobyl Nuclear Power Plant.

TABLE 17. DETAILED BREAKDOWN OF WASTES RESULTING FROM THE CHERNOBYL ACCIDENT AND THEIR LOCATION [5] (cont.)

RADIOACTIVE WASTE ASSOCIATED WITH THE SHELTER OBJECT			
Fuel-containing mass with a nuclear fuel content greater than 1%			
Radioactive Materials	RW type by dose rate or activity	Quantity	Mass (tonnes)
Concrete, concrete slabs and blocks	HLW ILW LLW	900 5 800 131 000	
Metal structures	LLW		1 440

RADIOACTIVE WASTE IN STORAGE AT CHORNOBYL NPP SITE

Solid radioactive waste			Liquid radioactive waste		
Type	Volume (m^3)	Activity (MBq)	Type	Volume (m^3)	Activity (MBq/m^3)
Low and intermediate level short lived waste	2 002	4.22E+6	Evaporation bottoms	13 581	28 175.4
High level waste	515	1.4E+8	Ion exchange resins	4 110	552.4
Low and intermediate level long lived waste	1.5	1.38E+5	Pulp	2 298	1,688
			Spent radioactive oil and oil–fuel mixture	145.3	0.181

RADIOACTIVE WASTE MANAGED WITHIN THE EXCLUSION ZONE

Radioactive waste disposal sites (RWDSs)

Facility	Volume (m^3)	Activity (Bq)	Status	Comment
Burjakovka	690 000	2.54E+15	Operational	Mainly LLW and ILW; however early trenches have some HLW
Pidlisny	3 960	2.59E+15	Operations ceased	Contains ILW and HLW in the form of FCMs
Chornobyl NPP Stage III	26 200	3.02E+14	Operations ceased	Contains LLW, ILW and HLW in one compartment
Vektor Complex	670	1.73E+11	Operational	Designed to dispose LLW and ILW
Total	720 380	5.47E+15		

TABLE 17. DETAILED BREAKDOWN OF WASTES RESULTING FROM THE CHERNOBYL ACCIDENT AND THEIR LOCATION[5] (cont.)

RADIOACTIVE WASTE ASSOCIATED WITH THE SHELTER OBJECT				
Fuel-containing mass with a nuclear fuel content greater than 1%				
Radioactive Materials			RW type by dose rate or activity	Quantity / Mass (tonnes)
Radioactive waste temporary storage places (RWTSPs)				
Facility	Volume (m^3)	Activity (Bq)	Status	Comment
New Stroybaza	21 950	6.61E+12	—	Contaminated topsoil, vegetation and construction debris
Old Stroybaza	40 150	3.52 E+13	—	Contaminated topsoil, vegetation and construction debris
Neftebasa	95 430	3.09E+13	—	Contaminated topsoil, dead pine trees, construction debris
Chistogalovka	874	7.11E+10	—	Wastes from the NPP industrial site and its surroundings
Yanov Station	30 000	3.70E+13	Partially investigated	Contaminated topsoil, dead pine trees and construction debris
Kopach	110 000	3.33E+13	Partially investigated	Construction and demolition waste
Red Forest	500 000	3.74E+14	Partially investigated	Contaminated trees, topsoil layer, forest litter and some building debris
Peschanoe Plato	57 288	5.31 E+12	—	Contaminated topsoil layer
Pripyat	16 000	2.59E+13	Partially investigated	Contaminated machinery, wood, construction waste, household waste, etc.
Total RWTSPs	871 692	1.99E+15	—	—

Note: RW: radioactive waste

III.3. INITIAL WASTE DISPOSAL ARRANGEMENTS

In 1986, the preferred method of waste disposal in the accident aftermath was disposal in situ, at the site where waste was formed (as discussed above for the NPP wastes). This was driven primarily by technical options for collection, packaging and transportation of the huge amounts of waste and the absence of nearby regular storage facilities capable of receiving them. In addition, the situation required immediate action. However, the localized facilities were regarded as temporary, with redisposal to facilities meeting longer term environmental safety requirements being considered. A committee was established to manage disposal and it decided on the location, design and urgent construction of disposal facilities near the villages of Burakovka and Pidlisny, within a few kilometres of the NPP.

During the emergency stage, 51 storage locations were developed in the exclusion zone as temporary stores for 135 500 m³ of wastes. These were constructed rapidly and their designs were extremely simple: 3–5 m deep trenches with a waterproof PVC liner — once filled, the liner edges were wrapped over the top of the waste and the trench was covered with clean soil. Trenches were sometimes located in ditches and gullies, and overfilling sometimes led to waste being above grade, in earth covered mounds. During 1987–1988, some of the wastes were removed to the radioactive waste disposal sites RWDS Burakovka and RWDS Pidlisny, but there are no surviving records of the material movements that took place. A famous temporary disposal site was known as the Red Forest, where a 200 hectare forest of Scots Pine that lay directly under the intense westerly plume containing core particulates had been killed by the deposited radioactivity. The trees were cut down and disposed of in situ, in soil covered trenches. Overall, while there are estimates, there is no reliable inventory of the contents of the wastes that were disposed of in the many locations used at and around the NPP in this early period [190]. A 2012 estimate is that there are ~886 400 m³ of waste in these temporary storage facilities [191].

The subsequent radioactive waste disposal site RWDS Buriakivka was designed along conventional lines for a near surface LLW repository (Fig. 59), with 30 clay lined trenches, each with a capacity of ~23 000 m³. The facility is considered to provide adequate isolation of the wastes in an area with no population.

RWDS Pidlisny, developed at the same time as RWDS Buriakivka, contains loose tipped or containerized HLW under a layer of concrete, set within concrete walls constructed on a concrete base on the land surface, with a total volume of ~11 000 m³. The facility is today covered by metal roofs. No safety assessment has been carried out on this facility. The RWDS Chornobyl NPP Stage III, on the NPP site was developed by making use of the unfinished (at the time of the accident) 'third stage' storage facility for LLW and ILW. It comprises seven concrete lined trenches, ~5 m deep and between 90 m and

FIG. 59. Radioactive waste disposal site Buriakivka. Courtesy of State Specialized Enterprise Centralized Enterprise for Radioactive Waste Management.

140 m long, with wastes emplaced both loose and in ~18 000 1 m³ metal containers, and covered with concrete slabs and earth.

III.4. LESSONS LEARNED IN THE EARLY STAGE WASTE MANAGEMENT PERIOD

In the early cleanup period, among many lessons learned on managing wastes, were:

(a) There was no immediately suitable equipment available for collecting and transporting wastes, especially those with high activity. Converted equipment did not perform well in the high radiation environment.
(b) There was no system for characterizing and segregating wastes effectively, or for material accounting. This lack of information has made present day management decisions difficult.
(c) There was no procedure to handle wastes in the emergency phase. Designs were lacking for easily constructed and operated temporary storage that would meet safety criteria and for quickly erected burial sites for long term storage of HLW.

III.5. CURRENT FACILITIES FOR WASTE STORAGE AND DISPOSAL

Since the 1980s, work has taken place to remove and consolidate wastes from many of the numerous temporary trenches. In particular, wastes in some locations where surface water contamination was occurring have been removed. Today, there remain nine temporary waste storage areas (RWTSPs) in the region within 10 km of the NPP site and the three RWDS, at Buriakivka, Pidlisny and Chornobyl NPP Stage III, developed shortly after the accident.

RWDS Buriakivka is still in use while new on-site facilities are being developed. Work to stabilize RWDS Pidlisny and install a metal roof took place in 2013. However, the Pidlisny is an accident legacy facility that is not suitable for the contained inventory. Knowledge of its barrier systems is limited or insufficient and it is expected that the contained inventory will be removed and transferred into an appropriate disposal site, when available. RWDS Chornobyl NPP Stage III is also not considered to be suitable for final disposal and the site remains unfinished, with engineered barriers that are considered inadequate. The facility is not equipped with water protection systems and the lower parts of the facility are periodically saturated with groundwater. In addition, gaps have developed in the earthen cover. Finally, RWDS Chornobyl NPP Stage III is located close to a surface water body, which is also considered not to be suitable for long term safety.

The Vektor Complex, located ~17 km from the NPP, comprises a number of facilities for handling and storing wastes, as well as several near surface disposal facilities: SRW1 for containerized wastes, SRW2 for bulk wastes and the Engineered Near Surface Disposal Facility (ENSDF, also referred to as Lot 3). These capabilities are being extended with new near surface disposal facilities and stores for HLW and spent fuel. It is expected that the Vektor Complex will eventually be able to receive most of decommissioning wastes from Units 1–3.

On the NPP site itself, a number of newer facilities are in operation, for example to receive wastes produced during foundation construction of the new NSC shelter. A considerable amount of liquid waste continues to be produced and the new facilities include a liquid radioactive waste treatment plant (LRTP), which was commissioned in 2018 to treat and treat and solidify ~20 000 m³ of accumulated liquid waste.

A legacy solid waste storage facility comprising an above ground concrete structure divided into three compartments with a total capacity of ~4000 m³ for storing all types of solid wastes (including HLW) was in operation from 1978 until 2003. A new solid waste retrieval facility is being developed as part of the new ICSRM at the site to remove wastes from the solid waste storage facility. The ICSRM project is one component of the integrated radioactive waste management programme for the entire NPP and includes waste retrieval, processing and packaging for interim storage or final disposal. This involves volume reduction through compaction and incineration and solidification through immobilization in concrete. There is also a plant for on-site manufacture of drums and concrete boxes to contain the wastes [192]. An interim temporary solid high level waste storage facility was commissioned in February 2004 to contain wastes until the ICSRM was completed. The ICSRM comprises a group of facilities constructed to modern international standards in waste management. The ICSRM facilities includes a temporary storage facility for low and intermediate level long lived and high level waste, a solid radwaste retrieval facility for designated for radioactive waste retrieved from existing Chornobyl NPP solid waste storages, and the Solid Radioactive Waste Treatment Plant (SRWTP).

Dismantling of the Unit 4 reactor building and the original Shelter Object will take place within the NSC structure and is expected to generate ~3350 drums of HLW and LILW-LL, which will be handled by the ICSRM, as well as a considerable amount of lower level wastes and further liquid wastes. Waste handling and generation activities are consequently expected to continue until at least 2065, as dismantling and remediation at the site reaches completion.

Appendix IV

THE FUKUSHIMA DAIICHI ACCIDENT

The M_w 9.0 Great Tohoku Earthquake of 11 March 2011 and the associated tsunami were an enormous natural disaster that caused the deaths of more than 18 000 people on the east coast of northern Honshu, Japan. The events overwhelmed the coastal Fukushima Daiichi NPP, causing flooding and prolonged total power loss, which led to the meltdown of three of the six reactors (Units 1–3) and hydrogen explosions that wrecked the buildings of Units 1, 3 and 4. This appendix provides a brief overview of the accident and of the subsequent work to manage the wastes that have resulted from the damage to the nuclear power station and the releases of radioactivity. A detailed account is provided by the IAEA [193], from which much of the information in this appendix was drawn.

IV.1. IMPACT OF THE EARTHQUAKE AND TSUNAMIS

At the time of the accident, three of the six reactor units at Fukushima Daiichi NPP were operating and three were on periodic closedown for refuelling or inspection. The earthquake triggered the automatic closedown of the operating reactors (Units 1–3), but resulted in the loss of external power from the grid to the nuclear power station. Shortly after the earthquake, a series of tsunamis struck the Pacific coast of Honshu. The second tsunami was 14 m high at Fukushima and overtopped the tsunami defence walls, flooding the ground floors and basements of the reactor buildings at the station and destroying the emergency diesel power generators of Units 1–4, which were located at low elevation. The impact of the tsunami removed any possibility of effective control of the reactors that had been operating, which began to overheat (decay heat). Eventually, this led to the meltdown of the cores of Units 1–3 and hydrogen gas explosions in Units 1, 3 and 4, which wrecked the reactor buildings and further affected the ability to work on and, eventually, clean up the site. Venting and damage to the reactor containment vessels led to airborne release of radioactivity and deposition, both offshore and onshore, along the coastal area around Fukushima and neighbouring prefectures. A decision was taken to evacuate the population in an area within 20 km of Fukushima Daiichi NPP.

Neither the accident at Fukushima Daiichi NPP nor the releases of radioactivity resulted in any deaths. Some 300 000 people were evacuated, of whom 160 000 were relocated for the long term, causing considerable stress that, it is argued, has led to hundreds of premature deaths. An area from which the population has been completely evacuated is undergoing remediation. It is surrounded by a larger area, where some evacuees have been allowed to return, which has also been subject to a major cleanup programme. Cleanup, remediation and decommissioning work will continue for some decades. The Fukushima Daiichi NPP will be decommissioned (and the undamaged reactors have not operated since the accident), although no decision has yet been reached on the end state of the site.

IV.2. WASTE MANAGEMENT

The emergency phase, leading to a state of cold shutdown of Fukushima Daiichi NPP, had been achieved around the end of 2011, at which stage the main phase of on-site stabilization and off-site remediation began. At the time, there was no law that regulated the disposal of accident waste and radioactive contaminated material, and the existing near surface waste disposal facility for radioactive waste from normal operation at the nuclear power station is not available for the disposal of radioactive contaminated materials. The wastes generated in the accident are summarized in Table 18.

TABLE 18. SOLID AND LIQUID WASTES RESULTING FROM THE FUKUSHIMA DAIICHI ACCIDENT

RADIOACTIVE WASTE MANAGED AT FUKUSHIMA DAIICHI NPP	
SOLID WASTE (TYPES AND QUANTITY)	LIQUID WASTE (TYPES AND QUANTITY)
Rubble: — <0.1 mSv/h: 268 200 m^3 — 0.1–1 mSv/h: 500 m^3 — Solid waste storage house: 21 100 m^3 Felled trees: 134 200 m^3 Discarded PPE: 48 200 m^3 Waste generated from water treatment: — Caesium adsorption vessel, ALPS spent vessel: 4 686 vessels — Sludge: 597 m	Liquid waste: — Condensed wastewater: 9 345 m^3 — Treated water (including Sr treated water): 1 201 434 m^3
FUKUSHIMA DAIICHI NPP OFF-SITE REMEDIATION	
Waste managed within Fukushima Prefecture	
Solid waste (types and quantity)	Liquid waste (types and quantity)
Soil from off-site decontamination activities: — ~14 million m^3 Waste within the Countermeasure Area: Transported to temporary storage sites: 2 630 000 t Incineration ash: — <100 000 Bq/kg: 1.55 Mm^3 — >100 000 Bq/kg: 20 000 m^3 Designated wastes (more than 8 000 Bg/kg): — Incinerator ash: 231 912 m^3 — Other: 28 700 m^3	No liquid wastes are expected
Waste managed outside Fukushima Prefecture	
Designated wastes (more than 8 000 Bg/kg): — Incinerator ash: 8 841 m^3 — Others: 18 732 m^3	No liquid wastes are expected

IV.2.1. Wastes from off-site remediation

Remediation activities have focused on intensive decontamination of residential areas to control external exposures, with control of internal exposures focused on restrictions and monitoring of foodstuff and animal feed, and the remediation of farmland. These activities have been the principal generators of radioactive wastes to date. The Japanese government has taken a conservative approach to establishing dose limits to people that has significantly increased the amount of remediation work that has been necessary, as well as affecting the movement and general treatment of the population. Remediation has been completed in several municipalities, with the return of the evacuated population being permitted.

A strategy was developed for managing the off-site wastes, involving:

(a) Collection in temporary storage sites near the decontamination location;
(b) Transport from temporary storage sites to an ISF;

(c) Volume reduction of combustible material by incineration in available municipal solid waste incinerators equipped with off-gas cleaning systems for the retention of ^{134}Cs and ^{137}Cs;
(d) Volume reduction of soil by washing to separate Cs or Cs rich constituents;
(e) Disposal, depending on radioactivity content, in the ISF or in commonly used or specially designated municipal landfills or near surface disposal facilities;
(f) Establishment of an inventory to keep track of activity and amounts accumulated.

Cleanup is complicated by the need also to deal with 'disaster wastes' from the impacts of the earthquake and tsunami in much of the area also affected by radioactive contamination. The total amount of solid contaminated waste that has had to be dealt with within the inner remediation area (the special decontamination area, or SDA) is estimated to be ~800 000 t. A further approximately 160 000 t of 'designated wastes' (with activity >8000 Bq/kg), largely comprising incineration ashes and sludges, is also present.

Surface soils and other materials have been dug up, with areas used by the public and children prioritized. Hot spots were identified by survey and included, in particular, areas where rainwater drains or is collected. Forests accumulate contamination, especially foliage and bark, and the burning of firewood concentrates contamination in the ash. To avoid the accumulation of highly contaminated ash, large amounts of unprocessed bark have been stored. Decontamination activities that generate wastes include the removal of soils and vegetation and the washing of structures.

Storage and disposal of the radioactively contaminated material generated following remediation has been a key challenge. Identifying temporary storage sites was challenging because of the concerns of landowners and constraints in gaining consent, as many owners were relocated to distant areas of Japan. Detailed arrangements have been made between the national, regional and local governments for how waste stores and disposal facilities are to be located and how wastes are to be routed to them in the different municipal areas in the zones being remediated. These arrangements are not discussed in this short appendix. As an interim measure, some material was stacked in covered piles in weatherproof sandbags or waterproof bags or containers near the decontaminated sites before being transported to specially constructed temporary storage sites.

At the point of collection, the segregation of contaminated material takes into account the waste form (combustible/non-combustible/soil) and the activity concentration of caesium (<8000 Bq/kg; <100 000 Bq/kg; >100 000 Bq/kg). Waste below 8000 Bq/kg is treated by normal methods used for non-radioactively contaminated waste (combustion, recycling of metals and plastic, composting organic materials, etc.) and is managed as municipal solid waste, utilizing the existing infrastructure for transportation, handling, volume reduction and disposal in municipal solid waste landfills. Designated waste (exceeding 8000 Bq/kg) requires special arrangements for transport, treatment, eventual recycling and reuse, or disposal in designated landfills equipped with systems for leachate collection, control of gases and monitoring.

Following temporary storage, wastes are to be shifted to the ISF. Various design types are considered for the ISF, depending on the activity of the wastes and the environment of the site, including lined trench facilities, with or without water management systems, and above grade concrete structures. A site for the ISF has been identified close to the Fukushima Daiichi NPP, with the facility expected to contain between 16 and 22 million cubic metres of waste — predominantly contaminated soils, after volume reduction, which will be held in a trench type facility. The much smaller volumes of higher activity (>100 000 Bq/kg) wastes will be stored in concrete buildings at the facility.

Designated wastes will eventually be disposed of in leachate controlled landfills, which will need to be sited and constructed in some areas.

Arrangements for the final disposal of off-site wastes have not been finalized. It is expected that the bulk of the soils in the ISF will remain in situ (with a mix of limited land reuse within, and controlled reuse after 30 years), while wastes with residual higher activity will be transported and disposed of outside Fukushima Prefecture.

FIG. 60. Decontamination and collection of soil in an agricultural area (left) and a temporary soil storage facility (right). Courtesy of Okuma Town.

IV.2.2. Wastes from on-site stabilization and decommissioning

Solid radioactive wastes were generated in the process of gaining access to and stabilizing the damaged nuclear facilities. Subsequently, significant quantities of waste were and are being generated during the implementation of the remediation programme. The initial wastes were principally rubble and other debris (concrete, metal, plastic) and felled trees and vegetation. The quantities are very much larger and levels of contamination much higher compared with waste originating from routine operations.

By mid-2015, the main pre-decommissioning cleanup work had been carried out and ~160 000 m^3 of debris and ~83 000 m^3 of felled trees were being stored on-site. The debris (from the damaged buildings and various cleanup activities on-site) is segregated based on the surface dose rates and types of material (e.g. concrete or steel) and stored in a range of structures, including soil covered heaps, tents, several hundred conventional transport containers and repurposed buildings (see Fig. 60). Rubble is stored in mounds with impermeable covers under a layer of soil to reduce radiation exposures. Felled trees (many of which were cut down to allow construction of contaminated water management facilities), after removal of vegetation, were stacked in heaps up to 5 m high, permitting air throughflow and watering to prevent fire.

The decommissioning roadmap comprises three stages, the first of which (removal of spent fuel from the storage ponds) was completed for Unit 4 in late 2014. The second stage involves preparations for the removal of fuel debris from the damaged reactors and will continue into the 2020s, with the final stage being site decommissioning, which will take a further 30 or 40 years. Decommissioning of the facility (removal of the structures, systems and components) cannot begin until fuel and fuel debris have been removed. In the final stage, removal of the fuel debris from the reactors is likely to take much longer than at TMI-2, where it was achieved in ~12 years. Up to 30 years has been allowed for in planning.

During decommissioning, by the end of the fuel removal cycle, it is conservatively estimated that approximately 560 000 m^3 of contaminated material will be generated. The planned new centralized storage facility will have a capacity of approximately 160 000 m^3. It is assumed that the difference between the expected amount of 560 000 m^3 and the planned amount will be addressed through proper segregation of the waste, volume reduction and recycling. A new incinerator unit for on-site wastes came into operation in March 2016.

IV.2.3. Liquid waste management

One of the major issues has been the management of large volumes of contaminated water, both from the continued requirement to circulate water to cool the damaged reactors and from leakage of surface and groundwater into the damaged reactor buildings. Approximately 800 m^3 of contaminated water is managed every day. Half the water is recirculated through the reactor containments to cool the fuel debris, with the other half being treated for decontamination. Allowing continued natural inflow to the damaged basements reduces the possibility of contamination migrating outwards into the ground. Approximately one million tonnes of treated water is stored on-site in approximately 900 tanks.

A variety of water treatment plants have been used to extract radioactive components from the waste, including a multiradionuclide removal facility: the ALPS. These facilities generate secondary wastes, such as sludges, slurry, adsorption vessels and adsorption media, which are held in concrete containers, stacked in temporary outdoor stores with concrete covers. Whilst removing most radionuclides, it is not possible to remove tritium, meaning that the water cannot currently be discharged. At the end of 2018, it was estimated that the planned storage capacity for treated water (1.37 million m^3) would be exceeded within three or four years [194]. Techniques are being evaluated to vitrify and reduce the volume of the secondary wastes.

The management of on-site waste poses complex R&D issues, many of which are unprecedented and challenging. There is cooperation among domestic and overseas organizations, and relevant experience and knowledge worldwide is being utilized.

Appendix V

OTHER NUCLEAR ACCIDENTS

This appendix briefly describes experience in managing wastes from the cleanup of a series of historic nuclear accidents in both the civil and military sectors.

V.1. KYSHTYM, FORMER USSR: 1957

The accident at the Mayak Production Association in September 1957 was caused by the explosion of a tank containing high activity waste solution from reprocessing at a plutonium production plant, near the town of Kyshtym, former USSR, ~100 km south of Sverdlovsk [176, 195]. It caused contamination of the River Techa and Lake Karachay. Approximately 10^{16} Bq of fission products were released, with the short lived radionuclides ^{144}Ce and ^{95}Zr accounting for 91% of the total released activity. After decay of the short lived radionuclides, the longer term radiological hazard was caused by ^{90}Sr, which amounted to 4×10^{15} Bq. Radioactive deposition took place in dry atmospheric conditions, contaminating an area of 15 000 km^2, with ^{90}Sr activity higher than 3.7 kBq/m^2 [195]. The radioactivity content of the releases was predominantly fission products, with very low contributions from activation products and actinides (e.g. Pu isotopes only accounted for 0.0043% of the activity).

There was a significant effort to segregate the waste according to the material and its origin, but due to the time pressure and the very large amounts of materials, extensive waste characterization and segregation of these materials was not possible. More than 6000 hectares of agricultural land were deep ploughed to bury the surface layer of heavily contaminated soil to a depth >50 cm, below the root penetration depth of many crops. Between 1958 and 1959, more than 20 000 hectares of land was ploughed to reduce uptake of contamination by plants and decrease gamma exposures. Less contaminated areas were decontaminated by moving the upper soil to lower lying terrain, such as abandoned construction pits and ditches. Approximately 320 000 m^3 of soil were removed and disposed. Remediation of the Mayak area is still ongoing, and it is clear that the actual site characterization process will be very difficult and take a long time.

The Kysthym accident is not the most serious contamination problem in the Mayak plant area. After stopping the dumping (from 1949 to 1956) of high level radioactive waste solutions directly into the River Techa, the HLW was then released into Lake Karachay. Today the contamination in Lake Karachay is $\sim 4.4 \times 10^{18}$ Bq [196] or a similar order of magnitude to what was released by the Chernobyl accident [197].

V.2. PALOMARES, SPAIN: 1966

In January 1966, two US military aircraft, one carrying four thermonuclear weapons, collided during an inflight fuel transfer. Three bombs were found on land the day after. Two of the weapons were seriously damaged by the detonation of a non-nuclear explosive device upon ground impact. The fourth was later recovered intact from the sea. The explosion spread plutonium oxide over ~660 hectares, contaminating soil, agricultural crops and other vegetation near the village of Palomares, which at that time had ~2000 inhabitants [198].

Soil with radioactive contamination levels above 1.2 MBq/m^2 was placed in 250 L drums and shipped to the Savannah River Plant in South Carolina, USA for burial. A total of 2.2 hectares was decontaminated by this technique, producing 6000 drums. Seventeen hectares of land with lower levels of contamination were mixed to a depth of 30 cm by harrowing and ploughing. On rocky slopes with contamination above 120 MBq/m², the soil was removed with hand tools and shipped to the USA in barrels [75].

Six zones were established to guide the initial remediation work, which was specified in an agreement between the Spanish and US governments. Cleanup involved burning crops with Pu levels less than 400 cpm. Crops above that limit were mulched, stored and later removed to the USA, along with 823 m^3 of soil with Pu levels exceeding 60 000 cpm. Approximately 115 hectares of soil with Pu levels ranging from 700 to 60 000 cpm were watered and ploughed. No remediation was performed in nearby hilly areas due to the steepness of the slopes.

The introduction of intensive farming in the 1980s, together with increased tourism in the 1990s, the construction of nearby water reservoirs and subsequent flooding, led to an expanded radiological surveillance programme and land use restrictions. This programme evaluated the residual source term and its distribution at depth through extensive characterization of a 660 hectare area, intensive characterization of 41 hectares and georadar mapping of pits. Ongoing environmental sampling includes air, soil, sediments, water, vegetation and animal products. Voluntary human bioassay sampling is underway to establish the general population's health, provide early detection of health conditions and lessen public fears. The public is also involved in the evaluation of further remediation options [172].

A 2004 survey revealed that there was still significant contamination present in certain areas and the Spanish government subsequently expropriated some plots of land, which would otherwise have been assigned for agricultural use or housing construction. In 2006, the governments of Spain and the USA agreed to decontaminate the remaining areas and share the workload and costs. In 2008 two trenches were discovered where the US Army had stored ~2000 m³ of contaminated earth during the 1966 operations. The trenches were found near where one of the nuclear devices was retrieved in 1966 and were probably dug at the last moment by US troops before leaving Palomares. In 2015 Spain and the USA signed an agreement to further discuss the cleanup and removal of land contaminated with radioactivity.

Lessons learned on waste management from the Palomares accident issues include:

- Changing, more intensive land uses (e.g. use of pesticides), as well as natural phenomena (e.g. floods) can require extensive later remediation and expanded land use restrictions;
- The initial soil removal, watering and ploughing measures were effective in diluting contamination into deeper soil horizons;
- A combination of wet and dry sieving proved effective in separating soil particles by size, minimizing the volume requiring disposal as radioactive waste.

V.3. CUIDAD JUAREZ, MEXICO: 1983

In December 1983, a teletherapy unit containing a sealed ^{60}Co radioactive source was dismantled and the source ruptured at a scrap yard in Ciudad Juarez, Mexico. The source contained over 6000 ^{60}Co pellets with a total of 1.7 × 1013 Bq, and these were scattered among the metal scrap. The scrap metal was subsequently used in local foundries, resulting in the production of thousands of tonnes of contaminated product. The contamination remained undetected and steel products were shipped both within Mexico and to the USA. In fact, the first recognition that an accident had occurred was when contaminated metal products were detected in the USA over a month later. Only then were the shipments traced back to the scrap yard in Ciudad.

The accident and subsequent dispersal resulted in thousands of tonnes of metal products, several foundries, streets and hundreds of houses being contaminated. In the USA several hundred tonnes of contaminated steel products were located. A large scale cleanup programme was initiated [199]. ^{60}Co pellets and contaminated metal product were collected and stored in appropriate concrete containers and steel barrels. In addition, significant quantities of contaminated soil and the debris from 814 impacted buildings, which were partially or fully demolished, were collected for disposal. In total, 21 000 m^3 of radioactive waste were collected under the cleanup programme [200].

V.4. GOIÂNIA, BRAZIL: 1987

In September 1987 two local residents found an abandoned radiotherapy unit at a former clinic scheduled for demolition. The unit contained a radioactive source with 5×10^{13} Bq of ^{137}Cs. The locals removed the rotating assembly containing the source, which they transported to a nearby house where one of them lived. In the following days the equipment was damaged and scraps of it were handled by numerous people. The contamination became widespread, including the courtyard of Goiânia's public health department, the Vigilancia Sanitaria, where source remnants were deposited.

In the initial response some early actions were taken to secure the source remnants at the Vigilancia Sanitaria, which were covered in concrete. Subsequently, a decision was taken to complete a comprehensive survey before taking any further action. In total, approximately 10 different sites became sufficiently contaminated that cleanup was necessary. Dose rates as high as 50–2000 mSv/h were detected. It soon became clear to the technical staff that large quantities of waste would be produced in the cleanup and a site would be needed in the Goiânia area to which the waste, properly packaged, could be transported and stored. A temporary storage area was chosen in a sparsely populated area ~20 km from Goiânia and 2.5 km from Abadia de Goias [201].

The cleanup eventually generated approximately 3500 m³ or 6000 t of ^{137}Cs contaminated waste. This waste was disposed in 1995 in two near surface repositories, constructed as large concrete vaults. The first repository was dedicated to very low level waste (40% in volume with ^{137}Cs activities <87 Bq/g) and the second to higher activity waste (40% in volume with ^{137}Cs activities >87 Bq/g). Studies continue to be conducted to assess the long term safety of the sites [202].

Some of the lessons learned concerning waste management were [201]:

(a) Soil profiling is valuable for determining the layer to be removed and avoiding the removal of large amounts of clean soil;
(b) The amount of material removed was seen by the local population as proportional to the risk to which they had been exposed, causing misunderstanding of waste management activities;
(c) Later construction processes and urban services brought buried contaminated material back to the surface;
(d) Lack of remediation criteria and site specific information led to the use of conservative approaches and the generation of unnecessary public stress, waste and costs;
(e) Involvement of local people, governmental and non-governmental organizations is key for good decisions on the strategies to be adopted;
(f) There is need for a policy and strategy for remediation of contaminated sites and the management of the waste, which can be included as part of a national waste management policy and strategy.

V.5. ACERINOX, SPAIN: 1998

In May 1998, a medical radiotherapy ^{137}Cs source with an activity of approximately 4.5 TBq was accidentally melted in an electric furnace at Acerinox, a scrap metal reprocessing plant in Los Barrios, Spain [203]. The ^{137}Cs released from the foundry created a radioactive cloud, which the Acerinox chimney detectors failed to detect, but was eventually detected in France, Italy, Switzerland, Germany and Austria. Some of the vapours were caught in a filter system, resulting in contamination of the 270 t of ashes already collected. The ash was removed and sent to two factories for processing as a part of their routine maintenance. One factory received 150 t, which was then used in a marsh stabilization process, increasing the mass of the contaminated material to 500 t and contaminating the marsh.

The maximum activity of the wastes before beginning the decontamination operations was approximately 2000 Bq/g. The waste produced was put into two types of containers. The ash was put into 1 m³ bags and metallic, plastic and paper wastes, etc. were put into 220 L drums. In total, ~2000 t of ash waste and 150 t of other waste are held at various sites with the intention of disposing of them at the El

Cabril Disposal Facility. Various treatment methodologies have been assessed to reduce the volumes of waste requiring disposal or to declassify them so that they can be disposed of as EW.

Appendix VI

CLEANUP OF LEGACY NUCLEAR SITES

This appendix briefly describes experience in managing large volumes of wastes from the cleanup of legacy nuclear sites. These are not accident wastes, but their volumes, the desired site end states and the approaches that have been taken are relevant to managing large volumes of wastes from nuclear accidents. In addition, this appendix includes closely related information on the French facilities for disposing of large volumes of VLLW from the dismantling of nuclear facilities.

VI.1. UNITED KINGDOM LEGACY CIVILIAN NUCLEAR POWER SITES

The world's first commercial nuclear power reactors at Calder Hall, Windscale (now part of the Sellafield site) went critical in 1956 and the site has been the focus of UK fuel cycle activities in the 60 years since that time. In 2004, the UK government established the Nuclear Decommissioning Authority, which was charged with cleaning up the UK's civilian nuclear legacy, which is spread across 17 sites. These sites include uranium enrichment and fuel manufacturing facilities, reactor research sites, nuclear power reactor sites and the Sellafield spent fuel reprocessing facilities.

The Nuclear Decommissioning Authority developed a strategy and plan for legacy cleanup that is updated at least every five years [204]. A key aspect of the strategy is hazard reduction, which prioritizes work based on the highest hazards first. Where the risks are unacceptable, urgent actions are taken to reduce them. Where the risk is less significant, technical, social and economic factors are taken into greater account. However, the focus remains on reducing risk and hazard as far as reasonably practicable.

An important aspect in cleanup planning is establishing an appropriate end state: the physical condition of a site after all remedial work is complete. This includes a definition of the amount of any residual contamination that may be left in the ground, or associated with any buildings or subsurface structures that remain in place.

Rather than apply a single, blanket cleanup target across all sites, end states are tailored to each site, taking into account likely reuse scenarios. Given the wide distribution of sites and the differing demand for land for redevelopment, cleanup targets are proportionate to what the next likely reuse of a site might be. In certain cases, this could mean leaving some contamination in situ and controlling risks by restricting access. In other cases, it could mean removing all residual contamination from a site to enable it to be completely released from regulatory control (delicensed), allowing unrestricted reuse. The level of effort required to achieve different end states and the amount of radioactive waste produced will therefore vary, as illustrated in Fig. 61.

Cleanup efforts involve very large volumes of waste, most of it lightly radioactively contaminated or activated. Much of this material will result from decommissioning nuclear facilities (producing steel, concrete, bricks, etc.) but also includes soils from remediation of contaminated land. The total estimated volume of radioactive waste from all sources is estimated to be 4 770 000 m^3 [205] comprising:

— VLLW: 2 720 000 m^3;
— LLW: 1 600 000 m^3;
— ILW: 449 000 m^3;
— HLW: 1 500 m^3.

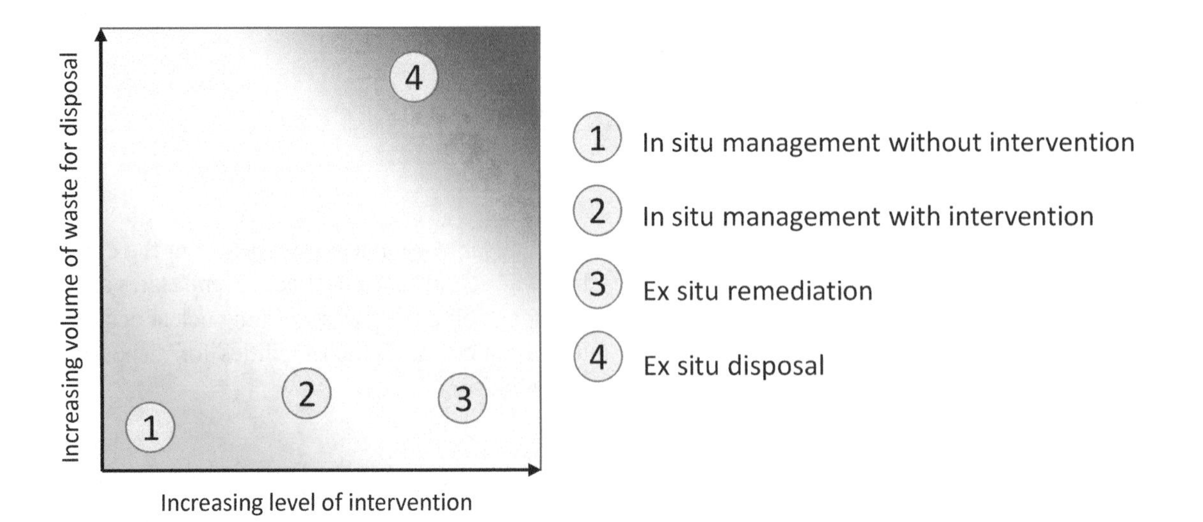

FIG. 61: Typical performance of options for remediating contaminated land on UK nuclear sites against key decision criteria. Adapted from Refs[205, 206].

Proportionate approaches to managing these large volumes of waste have to be followed to ensure efficient use of national resources. A revised national strategy for VLLW and LLW management was introduced in 2010 [207] and updated in 2016 [208]. This strategy promotes two key themes:

(a) Widespread application of the waste hierarchy, in particular to divert more materials to reuse and recycling routes, rather than to disposal;
(b) Better segregation of VLLW from LLW, and disposal of VLLW in near surface, landfill type facilities (rather than engineered vaults), based on appropriate risk assessment.

This strategy is supported by a revised clearance policy that allows waste materials to be managed (and disposed of) as 'out of scope' (i.e. non-radioactive) for regulatory purposes. The UK nuclear industry developed a practical guide and code of practice for applying clearance, which makes a useful distinction between clean materials that can be shown on the basis of provenance and records to have no potential to be contaminated or activated, and excluded materials that can be shown on the basis of measurement to contain radioactivity levels below clearance limits [209].

The combined application of clearance, segregation of VLLW and risk based disposal has substantially changed the way very large volumes of cleanup wastes are managed in the UK.

The UK operates one national LLW disposal facility: the LLWR, near Drigg, close to the Sellafield site. This surface engineered, vault type repository accepts LLW, which is cemented and containerized. The LLWR facility has a maximum capacity of 1.7 million m^3. Its operational life has been extended by diverting LLW to alternative landfills (now being developed commercially), in accordance with the national policy.

The overall legacy remediation lessons learned are that:

(1) Radioactive waste arising from accidents (e.g. the Windscale fire) can be managed as part of an integrated national waste management strategy that addresses all radioactive wastes;
(2) Cleanup needs to be planned using well defined and practical end state targets. The end state (hence cleanup levels) needs to be derived taking into account safety, technical and socioeconomic factors, and may vary from site to site;
(3) By taking a risk based, proportionate approach to disposal, better use of resources can be achieved. Diverting low risk wastes away from costly engineered facilities can preserve disposal capacity, facilitating timely disposal of other waste streams.

VI.2. UNITED STATES OF AMERICA: LEGACY NUCLEAR WEAPONS RESEARCH, DEVELOPMENT AND PRODUCTION SITES

Remediating thousands of contaminated locations at multiple sites across the USA that are or were once involved in the research, development and production of nuclear weapons is an ongoing, decades long task. It is estimated that more than 10 million m^3 of diverse wastes have been disposed of to date. Work is ongoing at many of these locations. Generally, new on-site disposal facilities have been built where sufficient waste volumes are present to achieve economies of scale. This approach also avoids the cost of loading, transporting and offloading waste at disposal facilities located hundreds or thousands of kilometres distant.

Local citizen advisory committees have been created for each of the cleanup sites to inform and involve local committees in remediation planning and implementation. As a result of this and the development of new disposal facilities in the vicinity of existing contaminated areas, local public opposition has been limited.

The Environmental Restoration Disposal Facility (ERDF) located on the Hanford Reservation in Washington state is one of the largest facilities developed for this purpose [210]. Opened in 1996, the site was developed to accept soil and debris containing low level radioactive waste, including material mixed with hazardous chemical constituents. The ERDF facility employs large, discrete disposal units with protective geomembrane underliner and drainage systems, and engineered covers. The design is shown in Fig. 62.

The ERDF currently includes eight disposal units, or cells. The newest disposal cells are 152 m wide, 305 m long and 21.3 m deep, with a capacity of 2.8 million t of waste. Economies of scale were further optimized through use of a single leachate collection system (previous designs employed two) and replacing a 30 cm thick gravel drainage layer with a geocomposite material, reducing capital costs and increasing waste capacity.

At small volume sites, or in the case of wastes requiring specialized disposal design and operating practices, centralized off-site disposal facilities are used for wastes from across the country. Examples of this include the ERDF, the near surface repository and deep bore hole disposal at the Nevada National Security Site (formerly the Nevada Test Site), and the Waste Isolation Pilot Project (WIPP) geological repository, which accepts transuranic waste from defence programmes. The Nevada National Security Site disposes of mixtures of radioactive and hazardous chemical waste and employs lined disposal trenches similar to those used at the ERDF, as well as borehole disposal.

The lessons learned from cleanup of legacy USA nuclear weapons sites include:

(a) Setting cleanup standards on a site specific basis allows greater flexibility in optimizing disposal solutions;
(b) Unit disposal costs were substantially reduced through the use of single purpose, very large volume disposal facilities;
(c) Locating disposal facilities near the site of waste generation offers significant transportation cost savings;
(d) For complex cleanups, multiple disposal facilities, designed, constructed and operated for different waste types, might be needed to optimize solutions;
(e) Realizing the benefits of this experience could be difficult in the case of a nuclear accident due to public pressure to implement more rapid solutions than this approach allows;
(f) Local citizens committees are an effective means to inform local communities about and involve them in cleanup planning and implementation.

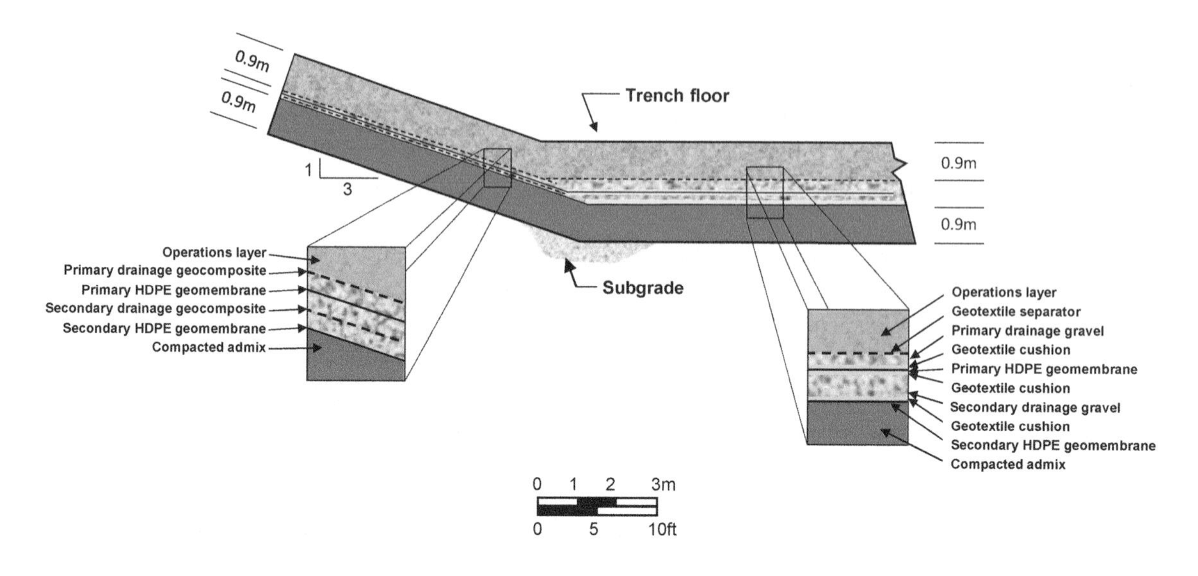

FIG. 62. Environmental Restoration Disposal Facility at Hanford. Courtesy of the US Department of Energy [210].

VI.3. AUSTRALIA: MARALINGA LEGACY NUCLEAR WEAPONS TEST SITE

Between 1952 and 1957, the United Kingdom conducted atmospheric nuclear weapons tests at three Australian sites, including the Maralinga Range. These tests and hundreds of minor trials produced long lived, highly radioactive contaminants. Some of these materials, ranging in size from inhalable dust to fragments, were ploughed under in 1967 to dilute concentrations and reduce human collection and environmental dispersal. The fragments included wire, rusted steel plate, lead, bitumen and yellow Bakelite. Unknown quantities of the highest risk materials were bulldozed into shallow pits 2–3 m deep. In 1979, 0.4 kg of Pu was removed from a burial pit and transported to the United Kingdom.

Contamination conditions drew the attention of Australian regulators in 1984, when large numbers of particles and fragments, highly contaminated with uranium, plutonium and americium, were discovered on and near the surface, over a wide area. Plutonium inhalation was the major hazard. The contaminated land earlier belonged to indigenous peoples who wished to resume ownership and use of the land. The Government responded by informing and involving these stakeholders while conducting land surveys and waste characterization studies.

Cleanup criteria were set based on a 5 mSv/a exposure limit. It was found to be economically unfeasible to clean up the entire area to this standard. Consequently, a 120 km^2 'non-residential area' was established, with human use limited to travel, hunting, camping and other transitory, non-residential activities. Alternative routes bypassing the area were also established, to limit use. Within this non-occupancy area, contaminated soils in discrete areas totalling 2.3 km^2 were removed, along with certain contaminated debris. Dispersed contamination was subject to soil removal to achieve levels <3 kBq/m^2 averaged over 1 hectare.

Remediation of the high activity burial pits involved in situ vitrification of half the pits, and exhumation and reburial of the remaining pit waste in a new, on-site, near surface disposal trench. In situ vitrification was halted following underground explosions caused by buried material. Approximately 263 000 m^3 of soil waste were disposed to a depth of 15 m and capped with a minimum of 5 m of clean soil. Land use restrictions are maintained at the disposal site to discourage intrusion and reduce the likelihood of exposure to any undiscovered fragments. Ongoing monitoring will address health and safety factors for visitors and users of the remediated areas [211].

The lessons learned from the cleanup of Maralinga include:

(a) The nature of the local environmental, land use and societal aspects guided waste disposal and other decisions;
(b) High dose rate scenarios were prioritized;
(c) At the time of contamination, the land was deemed worthless and uninhabitable forever: 40 years later, the land is highly valued by its owners, illustrating how values can change over a matter of decades;
(d) Disposal of uncharacterized waste in shallow pits made later cleanup difficult and costly;
(e) Combining stakeholder consultation, scientific expertise and engineering and technical support excellence enhanced public trust and confidence;
(f) Incorporating elements of reversibility and retrieval in disposal plans provided flexibility in addressing long lived wastes;
(g) In situ vitrification, without reasonable knowledge of buried waste constituents, invites problems.

The cleanup experience of legacy nuclear waste sites is summarized in Table 19.

VI.4. FRANCE: DISPOSING OF VERY LOW LEVEL RADIOACTIVE WASTE FROM DISMANTLING OF NUCLEAR FACILITIES

The CIRES very low level radioactive waste disposal facility at Morvilliers in France opened in 2003. It was decided to create such a facility in view of the absence of an unconditional clearance level for dismantling nuclear facilities in France, meaning that a significant volume of the materials from dismantling could not be reused and had to be disposed of as waste, even with very low contamination levels [212].

The basic principle of the repository design was to comply with regulations governing disposal facilities for non-radioactive hazardous waste. By applying such a principle, it is possible to accommodate both radioactive waste and toxic chemicals. Containment relies on identifying a low permeability surface clay layer in which the repository is developed.

TABLE 19. SUMMARY OF LEGACY NUCLEAR WASTE SITE CLEANUP

Experience	Waste generated
Legacy civilian nuclear site cleanup (United Kingdom)	Estimated 4 490 000 m^3 from all sources: mostly from decommissioning of nuclear facilities but includes soils from remediation of contaminated land: — 2 840 000 m^3 (63.2%) VLLW — 1 370 000 m^3 (30.5%) LLW — 286 000 m^3 (6.4%) ILW — 1 080 m^3 (0.1%) HLW
Legacy nuclear weapons site remediation (USA)	Estimated 10 million m^3 of diverse wastes disposed to date: — Work ongoing at several sites, with on-site disposal facilities built as needed where sufficient waste volumes are present to achieve economies of scale
Maralinga nuclear weapons test site cleanup (Australia)	120 km^2 non-residential area established, with human use limited to travel, hunting, camping and other transitory, non-residential activities: — Total 2.3 km^2 of contaminated soils in discrete areas removed with debris Remediation of high activity burial pits: — 263 000 m^3 of soil disposed to a depth of 15 m and capped with minimum 5 m of clean soil — In situ vitrification of half of high activity burial pits

As shown in Fig. 63, disposal trenches are excavated within a natural clay layer, with the trench sides and bottom being further protected by an impermeable geomembrane. Trenches are developed 8 m below grade and rise 4 m above grade at completion. The waste is placed over the membrane, while a mobile roof directs rainfall to drainage systems, protecting operations throughout loading. Containerized and certain bulk material is accepted. Filled trenches are backfilled and sealed with the same membrane. The repository is ultimately covered with clay and revegetated. An inspection hole is used to verify that there is no water seepage around the waste.

The CIRES facility has performed well during its 10 years of operation, with more than 30 000 m³ of waste being disposed of annually. More than 10 trenches have been excavated and filled with waste and the site capacity is sufficient to accommodate the waste being delivered. A final cover is implemented as trenches are filled with waste. The facility is expected to be operational for more than 20 years with a planned disposal capacity of 650 000 m³. The average activity content of the waste is ~9 Bq/g.

In January 2014, a new type of shelter called Prémorails (see Fig. 64) was installed. Easier to move and safer to operate, this effective patented device was developed by the waste management agency (Andra) to protect future vaults. It consists of a metal structure that can be divided into sections and is mounted on rails, with a tarpaulin cover. Innovation focuses on three aspects:

(a) Autonomy: the flexible and divisible structure moves by sections mounted on rails, rather than through the use of large lifting equipment, as was previously the case;
(b) Sealing: together, the sections form a 180 m long watertight shelter, with additional protective tarpaulins added between each assembly;
(c) Safety: the presence of a central passageway in the inside top of the shelter contributes to the safe assembly and disassembly of the different sections.

The safety approach at the Morvilliers facility is consistent with that adopted for the low level radioactive waste disposal facility at Centre de l'Aube, located 1.5 km from the CIRES facility, and addresses the impact of the facility with regard to both the radiological and the chemical toxicities of the waste. The toxic risk was established in relation to a threshold or reference value for As, Zn, Pb and Cd, and for chemical elements with carcinogenic effects (As, Cd). The radiological risk is characterized by dose calculations. Doses are compared with limits set by the regulatory agency, consistent with international standards.

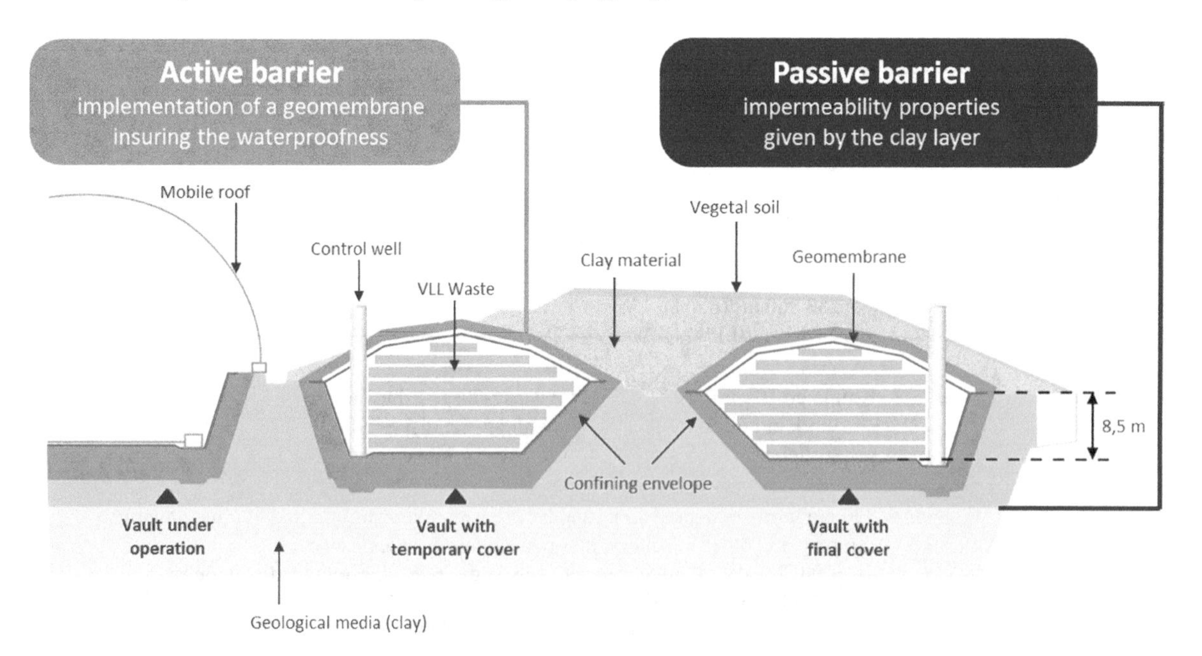

FIG. 63. CIRES Morvilliers VLLW disposal facility design. Courtesy of Andra.

FIG. 64. The Prémorails shelters at the CIRES facility.

REFERENCES

[1] DELTETE, P., HAHN, R., TMI-2 Waste Management Experience, EPRI TR-100640, Electric Power Research Institute, Palo Alto, CA, 1992.

[2] HOLTON, W.C., NEGIN, C.A., OWRUTSKY, S.L., The Cleanup of Three Mile Island Unit 2 — A Technical History: 1979–1990, EPRI NP-6931, Electric Power Research Institute, Palo Alto, CA, 1990.

[3] INTERNATIONAL ATOMIC ENERGY AGENCY, Environmental Consequences of the Chernobyl Accident and their Remediation: Twenty Years of Experience, Radiological Assessment Reports Series No. 8, IAEA, Vienna (2006).

[4] INTERNATIONAL ATOMIC ENERGY AGENCY, Chernobyl: Looking Back to Go Forward, Proceedings Series, IAEA, Vienna (2008).

[5] INTERNATIONAL ATOMIC ENERGY AGENCY, Present and Future Environmental Impact of the Chernobyl Accident, IAEA -TECDOC-1240, IAEA, Vienna (2001).

[6] INTERNATIONAL ATOMIC ENERGY AGENCY, One Decade after Chernobyl: Summing Up the Consequences of the Accident, Proceedings Series, IAEA, Vienna (1996).

[7] TOKYO ELECTRIC POWER COMPANY, Situation of Storage and Treatment of Accumulated Water Including Highly Concentrated Radioactive Materials at Fukushima Daiichi Nuclear Power Station (155th Release), Press Release (2014), https://www.tepco.co.jp/en/press/corp-com/release/2014/1237952_5892.html

[8] COMITÉ DIRECTEUR POUR LA GESTION DE LA PHASE POST-ACCIDENTELLE, Policies and Strategies for Radioactive Waste Management, Policy Elements for Post-Accident Management in the Event of Nuclear Accident, CODIRPA, Montrouge (2012).

[9] INTERNATIONAL ATOMIC ENERGY AGENCY, The Follow-up IAEA International Mission on Remediation of Large Contaminated Areas Off-Site the Fukushima Daiichi NPP, Final Report NE/NEFW/2013, IAEA, Vienna, 2014.

[10] INTERNATIONAL ATOMIC ENERGY AGENCY, The International Mission on Remediation of Large Contaminated Areas Off-Site the Fukushima Daiichi NPP, Final Report NE/NEFW/2011, IAEA, Vienna, 2011.

[11] INTERNATIONAL ATOMIC ENERGY AGENCY, IAEA International Peer Review Mission on Mid-and-Long-Term Roadmap Towards the Decommissioning of TEPCO's Fukushima Daiichi Nuclear Power Station Units 1–4, Tokyo and Fukushima Prefecture, Japan, 25 November–4 December 2013, Mission Report, IAEA, Vienna, 2014.

[12] INTERNATIONAL ATOMIC ENERGY AGENCY, IAEA International Peer Review Mission on Mid-and-Long-Term Roadmap Towards the Decommissioning OF TEPCO'S Fukushima Daiichi Nuclear Power Station Units 1–4, Tokyo and Fukushima Prefecture, Japan, 15–22 April 2013, Mission Report, IAEA, Vienna, 2013.

[13] INTERNATIONAL ATOMIC ENERGY AGENCY, Classification of Radioactive Waste, IAEA Safety Standards Series No. GSG-1, IAEA, Vienna (2009).

[14] FOOD AND AGRICULTURE ORGANIZATION OF THE UNITED NATIONS, INTERNATIONAL ATOMIC ENERGY AGENCY, INTERNATIONAL LABOUR OFFICE, PAN AMERICAN HEALTH ORGANIZATION, WORLD HEALTH ORGANIZATION, Criteria for Use in Preparedness and Response for a Nuclear or Radiological Emergency, IAEA Safety Standards Series No. GSG-2, IAEA, Vienna (2011).

[15] INTERNATIONAL ATOMIC ENERGY AGENCY, Predisposal Management of Radioactive Waste, IAEA Safety Standards Series No. GSR Part 5, IAEA, Vienna (2009).

[16] EUROPEAN COMMISSION, FOOD AND AGRICULTURE ORGANIZATION OF THE UNITED NATIONS, INTERNATIONAL ATOMIC ENERGY AGENCY, INTERNATIONAL LABOUR ORGANIZATION, OECD NUCLEAR ENERGY AGENCY, PAN AMERICAN HEALTH ORGANIZATION, UNITED NATIONS ENVIRONMENT PROGRAMME, WORLD HEALTH ORGANIZATION, Radiation Protection and Safety of Radiation Sources: International Basic Safety Standards, IAEA Safety Standards Series No. GSR Part 3, IAEA, Vienna (2014).

[17] INTERNATIONAL ATOMIC ENERGY AGENCY, Predisposal Management of Radioactive Waste from NPP s and Research Reactors, IAEA Safety Standards Series No. SSG-40, IAEA, Vienna (2016).

[18] INTERNATIONAL ATOMIC ENERGY AGENCY, Predisposal Management of Radioactive Waste from Nuclear Fuel Cycle Facilities, IAEA Safety Standards Series No. SSG-41, IAEA, Vienna (2016).

[19] INTERNATIONAL ATOMIC ENERGY AGENCY, Remediation Process for Areas Affected by Past Activities and Accidents, IAEA Safety Standards Series No. WS-G-3.1, IAEA, Vienna (2007).

[20] INTERNATIONAL ATOMIC ENERGY AGENCY, Storage of Radioactive Waste, IAEA Safety Standards Series No. WS-G-6.1, IAEA, Vienna (2006).

[21] INTERNATIONAL ATOMIC ENERGY AGENCY, Management of Large Volumes of Radioactive Waste Arising from Nuclear or Radiological Incidents, IAEA-TECDOC-1826, IAEA, Vienna (2017).
[22] INTERNATIONAL ATOMIC ENERGY AGENCY, Experiences and Lessons Learned Worldwide in the Cleanup and Decommissioning of Nuclear Facilities in the Aftermath of Accidents, IAEA Nuclear Energy Series No. NW-T-2.7, IAEA, Vienna (2014).
[23] INTERNATIONAL ATOMIC ENERGY AGENCY, Decommissioning after a Nuclear Accident: Approaches, Techniques, Practices and Implementation Considerations, IAEA Nuclear Energy Series No. NW-T-2.10, IAEA, Vienna (2019).
[24] INTERNATIONAL ATOMIC ENERGY AGENCY, IAEA Safety Glossary: 2018 Edition, Non-serial Publications, IAEA, Vienna (2018).
[25] RENO, H.W., SCHMITT, R.C., Historical Summary of the Fuel and Waste Handling and Disposition Activities of the TMI-2 Information and Examination Programme (1980–1988), EGG-2529, EG&G Idaho, Idaho Falls, ID (1988).
[26] INTERNATIONAL ATOMIC ENERGY AGENCY, Design Principles and Approaches for Radioactive Waste Repositories, IAEA Nuclear Energy Series No. NW-T-1.26, IAEA, Vienna (2020).
[27] INTERNATIONAL ATOMIC ENERGY AGENCY, Policy and Strategies for Environmental Remediation, IAEA Nuclear Energy Series No. NW-G-3.1, IAEA, Vienna (2015).
[28] INTERNATIONAL ATOMIC ENERGY AGENCY, Guidelines for Remediation Strategies to Reduce the Radiological Consequences of Environmental Contamination, Technical Report Series No. 475, IAEA, Vienna (2012).
[29] NUCLEAR REGULATORY COMMISSION, Licensing Requirements for Land Disposal of Radioactive Waste, 10 CFR 61, US Govt Printing Office, Washington, DC (1983).
[30] NUCLEAR REGULATORY COMMISSION, Programmatic Environmental Impact Statement Related to Decontamination and Disposal of Radioactive Wastes Resulting from 28 March 1979 Accident Three Mile Island Nuclear Station, Unit 2, NUREG 0683 Supplement 1, Docket No 50-320, US Govt Printing Office, Washington, DC (1984).
[31] INTERNATIONAL ATOMIC ENERGY AGENCY, Approach to Develop Waste Acceptance Criteria for Low and Intermediate Level Waste, IAEA, Vienna (in preparation).
[32] MINISTRY OF THE ENVIRONMENT, Recycling Technology Development Strategy Study Group: Volume Reduction and Recycling for Intermediate Storage Removal, Technical Issues on Recycling, Presentation (2015),
http://josen.env.go.jp/chukanchozou/facility/effort/investigative_commission/pdf/proceedings_150721_06.pdf
[33] EAST JAPAN GREAT EARTHQUAKE RESPONSE RESEARCH COMMITTEE, Characterization Scheme for Recycled Materials from Incineration Bottom Ash of Disaster Wastes for Ground Materials, Fact Sheet (2012),
http://geotech.gee.kyoto-u.ac.jp/JGS/Scheme_abstract01.pdf
[34] JAPAN CONSTRUCTION FEDERATION ASSOCIATION, List of Quality Standards Relating to the Reconstruction and Utilization of Disaster Waste, Fact Sheet (2012), www.nikkenren.com/pdf/disaster/2012_1109saigaihaiki.pdf
[35] INTERNATIONAL ATOMIC ENERGY AGENCY, Monitoring for Compliance with Exemption and Clearance Levels, Safety Reports Series No. 67, IAEA, Vienna (2012).
[36] INTERNATIONAL ATOMIC ENERGY AGENCY, Application of the Concepts of Exclusion, Exemption and Clearance, IAEA Safety Standards Series No. RS-G-1.7, IAEA, Vienna (2004).
[37] INTERNATIONAL ATOMIC ENERGY AGENCY, Derivation of Activity Concentration Values for Exclusion, Exemption and Clearance, Safety Reports Series No. 44, IAEA, Vienna (2005).
[38] INTERNATIONAL ATOMIC ENERGY AGENCY, Regulatory Control of Radioactive Discharges to the Environment, IAEA Safety Standards Series No. GSG-9, IAEA, Vienna (2018).
[39] INTERNATIONAL ATOMIC ENERGY AGENCY, Setting Authorized Limits for Radioactive Discharges: Practical Issues to Consider, IAEA-TECDOC-1638, IAEA, Vienna (2010).
[40] TYROR, J.G., PEARSON, G.W., "The medical implications of NPP accidents", Medical Response to Effects of Ionizing Radiation (Proc. Conf. London 1989) UKAEA, London (1989).
[41] CHAMBERLAIN, A.C., Emission of Fission Products and Other Activities During the Accident to Windscale Pile No. 1 in October 1957, AERE-M3194, UKAEA, London, 1981.
[42] NUCLEAR REGULATORY COMMISSION, Backgrounder on the Three Mile Island Accident, Fact Sheet (2018),
https://www.nrc.gov/reading-rm/doc-collections/fact-sheets/3mile-isle.html
[43] INTERNATIONAL ATOMIC ENERGY AGENCY, Regulatory Control for the Safe Transport of Naturally Occurring Radioactive Material (NORM), IAEA-TECDOC-1728, IAEA, Vienna (2013).

[44] INTERNATIONAL ATOMIC ENERGY AGENCY, Advisory Material for the IAEA Regulations for the Safe Transport of Radioactive Material (2012 Edition), IAEA Safety Standards Series No. SSG-26, IAEA, Vienna (2014).
[45] FOOD AND AGRICULTURE ORGANIZATION OF THE UNITED NATIONS, INTERNATIONAL ATOMIC ENERGY AGENCY, INTERNATIONAL CIVIL AVIATION ORGANIZATION, INTERNATIONAL LABOUR ORGANIZATION, INTERNATIONAL MARITIME ORGANIZATION, INTERPOL, OECD NUCLEAR ENERGY AGENCY, PAN AMERICAN HEALTH ORGANIZATION, PREPARATORY COMMISSION FOR THE COMPREHENSIVE NUCLEAR-TEST-BAN TREATY ORGANIZATION, UNITED NATIONS ENVIRONMENT PROGRAMME, UNITED NATIONS OFFICE FOR THE COORDINATION OF HUMANITARIAN AFFAIRS, WORLD HEALTH ORGANIZATION, WORLD METEOROLOGICAL ORGANIZATION, Preparedness and Response for a Nuclear or Radiological Emergency, IAEA Safety Standards Series No. GSR Part 7, IAEA, Vienna (2015).
[46] INTERNATIONAL ATOMIC ENERGY AGENCY, Leadership, Management and Culture for Safety in Radioactive Waste Management, IAEA Safety Standards Series No. GSG-16, IAEA, Vienna (2022).
[47] ORGANISATION FOR ECONOMIC CO-OPERATION AND DEVELOPMENT NUCLEAR ENERGY AGENCY, Management of Radioactive Waste after a NPP Accident, NEA Report 7305, OECD, Paris (2016).
[48] INTERNATIONAL ATOMIC ENERGY AGENCY, Disposal of Waste from the Cleanup of Large Areas Contaminated as a Result of a Nuclear Accident, Technical Reports Series No. 330, IAEA, Vienna (1992).
[49] INTERNATIONAL ATOMIC ENERGY AGENCY, Management of Abnormal Radioactive Wastes at NPPs, Technical Reports Series No. 307, IAEA, Vienna (1990).
[50] GENERAL ACCOUNTING OFFICE, Impact of Federal R&D funding on Three Mile Island Cleanup Costs, EMD-82-28, US GAO, Gaithersburg, MD (1982).
[51] INTERNATIONAL ATOMIC ENERGY AGENCY, Selection of Technical Solutions for the Management of Radioactive Waste, IAEA-TECDOC-1817, IAEA, Vienna (2017).
[52] INTERNATIONAL ATOMIC ENERGY AGENCY, Restoration of Environments Affected by Residues from Radiological Accidents: Approaches to Decision Making, IAEA-TECDOC-1131, IAEA, Vienna (2000).
[53] INTERNATIONAL ATOMIC ENERGY AGENCY, Maintenance of Records for Radioactive Waste Disposal, IAEA-TECDOC-1097, IAEA, Vienna (1999).
[54] INTERNATIONAL ATOMIC ENERGY AGENCY, Retrieval, Restoration and Maintenance of Old Radioactive Waste Inventory Records, IAEA-TECDOC-1548, IAEA, Vienna (2007).
[55] INTERNATIONAL ATOMIC ENERGY AGENCY, Waste Inventory Record Keeping Systems (WIRKS) for the Management and Disposal of Radioactive Waste, IAEA-TECDOC-1222, IAEA, Vienna (2001).
[56] NRC NEWS, NRC Approves License Transfer for Three Mile Island, Unit 2, Press Release (2020), https://www.nrc.gov/reading-rm/doc-collections/news/2020/20-058.pdf
[57] BALOGA V. I. et al., 20 Years After Chornobyl Catastrophe: Future Outlook, National Report of Ukraine, Atika, Kiev (2006).
[58] THE NUCLEAR DAMAGE COMPENSATION FACILITATION CORPORATION, The Nuclear Damage Compensation and Decommissioning Facilitation Corporation (NDF), Fact Sheet (2021), https://www.ndf.go.jp/files/user/soshiki/pamph_e.pdf
[59] TOKYO ELECTRIC POWER COMPANY, Mid- and Long-Term Roadmap Towards the Decommissioning of TEPCO's Fukushima Daiichi Nuclear Power Station, Report (2019), https://www.meti.go.jp/english/earthquake/nuclear/decommissioning/pdf/20191227_3.pdf
[60] ARNOLD, L., Windscale 1957: Anatomy of a Nuclear Accident, 3rd edn, Palgrave Macmillan, London (2007).
[61] MINISTRY OF THE ENVIRONMENT, Decontamination Guidelines, Report (2013), http://josen.env.go.jp/en/framework/pdf/decontamination_guidelines_2nd.pdf
[62] INTERNATIONAL ATOMIC ENERGY AGENCY, Mobile Processing Systems for Radioactive Waste Management, IAEA Nuclear Energy Series No. NW-T-1.8, IAEA, Vienna (2014).
[63] BONDARKOV, M. (Ed.), 30 Years of the Chernobyl Disaster (Reviews), KIM Publishing House, Kyiv (2016).
[64] INTERNATIONAL NUCLEAR SAFETY ADVISORY GROUP, Summary Report on the Post-accident Review Meeting on the Chernobyl Accident, INSAG Series No. 1, IAEA, Vienna (1986).
[65] INTERNATIONAL NUCLEAR SAFETY ADVISORY GROUP, The Chernobyl Accident: Updating of INSAG-1, INSAG Series No. 7, IAEA, Vienna (1993).
[66] EG AND G IDAHO INC., ALLIED-GENERAL NUCLEAR SERVICES, ARGONNE NATIONAL LABORATORY, ELECTRIC POWER RESEARCH INSTITUE, SANDIA NATIONAL LABORATORIES, Planning Report, GEND-001, Three Mile Island Operations Office, Washington DC (1980).
[67] FOOD AND AGRICULTURE ORGANIZATION OF THE UNITED NATIONS, INTERNATIONAL ATOMIC

ENERGY AGENCY, INTERNATIONAL CIVIL AVIATION ORGANIZATION, INTERPOL, PREPARATORY COMMISSION FOR THE COMPREHENSIVE NUCLEAR-TEST-BAN TREATY ORGANIZATION AND UNITED NATIONS OFFICE FOR OUTER SPACE AFFAIRS, Arrangements for Public Communication in Preparedness and Response for a Nuclear or Radiological Emergency, IAEA Safety Standards Series No. GSG-14, IAEA, Vienna (2020).

[68] TOKYO ELECTRIC POWER COMPANY, Reviewing the Two Years of Nuclear Safety Reform, Press Release (2014),
https://www4.tepco.co.jp/en/press/corp-com/release/betu15_e/images/150330e0302.pdf

[69] MINISTRY OF THE ENVIRONMENT, Decontamination Projects for Radioactive Contamination Discharged by Tokyo Electric Power Company Fukushima Daiichi Nuclear Power Station Accident, Internal Report (2019),
http://josen.env.go.jp/en/policy_document/

[70] INTERNATIONAL ATOMIC ENERGY AGENCY, Communication and Stakeholder Involvement in Radioactive Waste Disposal, IAEA Nuclear Energy Series No. NW-T-1.16, IAEA, Vienna (2022).

[71] OECD NUCLEAR ENERGY AGENCY, International Roundtable on the Final Disposal of High-Level Radioactive Waste and Spent Fuel, Summary Report NEA No. 7529, OECD, Paris, 2020.

[72] INTERNATIONAL ATOMIC ENERGY AGENCY, Policies and Strategies for Radioactive Waste Management, IAEA Nuclear Energy Series No. NW-G-1.1, IAEA, Vienna (2009).

[73] INTERNATIONAL ATOMIC ENERGY AGENCY, Arrangements for the Termination of a Nuclear or Radiological Emergency, IAEA Safety Standards Series No. GSG-11, IAEA, Vienna (2018).

[74] UNITED STATES ENVIRONMENTAL PROTECTION AGENCY, Technologies to Improve Efficiency of Waste Management and Cleanup After a Radiological Dispersal Device Incident: Standard Operational Guideline, EPA/600/R-13/124, US Environmental Protection Agency, Washington, DC (2013).

[75] LLERANDI, C.S., "Palomares: From the accident to the rehabilitation plan", Paper No. 2.3, paper presented at Int. Symp. on Decontamination: Towards the Recovery of the Environment, Fukushima, 2011.

[76] OUZOUNIAN, G., DUTZER, M., TORRES, P., Disposal of short-lived waste in France, News article (2012),
https://www.neimagazine.com/features/featuredisposal-of-short-lived-waste-in-france/

[77] MOLITOR, N., DRACE, Z., JAVELLE, C., "Achievements and remaining challenges for the conversion of Chornobyl NPP Unit 4 into ecologically safe conditions", 30 Years of the Chernobyl Disaster (Reviews), KIM Publishing House, Kyiv (2016).

[78] EUROPEAN COMMISSION JOINT RESEARCH CENTRE, Chernobyl Unit 4 — Follow-up Actions, Short and Long Term Measures for Chernobyl NPP, Contracts (2022),
https://nuclear.jrc.ec.europa.eu/tipins/contracts/chernobyl-unit-4-follow-actions-short-and-long-term-measures-chernobyl-npp

[79] NUCLEAR REGULATORY AUTHORITY, Order from NRA to TEPCO, Document No. 121107002, Correspondence (2012),
https://www.nsr.go.jp/data/000069063.pdf

[80] TOKYO ELECTRIC POWER COMPANY, Implementation Plan of the Measures for the Specified Reactor Facilities at Fukushima Daiichi Nuclear Power Station, Press Release (2013),
https://www.tepco.co.jp/en/press/corp-com/release/2013/1228248_5130.html

[81] MINISTRY OF THE ENVIRONMENT, Interim Storage Facility, Fact Sheet (2022),
http://josen.env.go.jp/en/storage/

[82] NUCLEAR REGULATORY COMMISSION, Guidance for the Reviews of Proposed Disposal Procedures and Transfers of Radioactive Material Under 10 CFR 20.2002 and 10 CFR 40.13(A), ML19295F109, US Govt Printing Office, Washington DC (2020).

[83] INTERNATIONAL ATOMIC ENERGY AGENCY Redevelopment and Reuse of Nuclear Facilities and Sites: Case Histories and Lessons Learned, IAEA Nuclear Energy Series No. NW-T-2.2, IAEA, Vienna (2011).

[84] MINISTRY OF THE ENVIRONMENT, On Progress of Decontamination, Presentation (2015),
http://josen.env.go.jp/material/session/pdf/015/mat05.pdf

[85] INTERNATIONAL ATOMIC ENERGY AGENCY, Application of Configuration Management in NPPs, Safety Reports Series No. 65, IAEA, Vienna, (2010).

[86] UNITED STATES ENVIRONMENTAL PROTECTION AGENCY, Guidance on Systematic Planning Using the Data Quality Objectives Process, EPA QA/G-4, US Environmental Protection Agency, Washington DC (2006).

[87] AREVA, AREVA Develops Two Innovative Solutions for Contamination Monitoring, Press Release (2012),
https://www.sa.areva.com/EN/news-9623/japan-areva-develops-two-innovative-solutions-for-contamination-monitoring.html

[88] MINISTRY OF THE ENVIRONMENT, Liaison and Coordination Meeting Regarding Transportation of Removed Soil to Intermediate Storage Facilities, Meeting Minutes (2020), http://josen.env.go.jp/chukanchozou/action/investigative_commission/pdf/transportation_200129.pdf
[89] MINISTRY OF THE ENVIRONMENT, Mid-and-Long-Term Roadmap Towards the Decommissioning of TEPCO's Fukushima Daiichi Nuclear Power Station Units 1–4, Monthly Progress Report (2020), https://www.meti.go.jp/english/earthquake/nuclear/decommissioning/
[90] MIRION TECHNOLOGIES, In-Situ Measurements: FoodScreen™ Radiological Food Screening System, Data Sheet (2019), https://mirion.s3.amazonaws.com/cms4_mirion/files/pdf/spec-sheets/c39444_foodscreen_spec_sheet_2.pdf?1562600743
[91] DUNSTER, H.J., HOWELLS, H., TEMPLETON, W.L., District surveys following the Windscale incident, October 19 J. Radiol. Prot. 27 (2007) 217–230.
[92] UNITED STATES DEPARTMENT OF ENERGY, In Situ Underwater Gamma Spectroscopy System, Innovative Technology Summary Report OST/TMS ID 2990, USDOE, Idaho Falls, ID (2001).
[93] AREVA, Mapping Contamination at Fukushima with Robots, News Brief (2013), https://web.archive.org/web/20141115193411/http:/areva.com/EN/news-9866/japan-mapping-contamination-at-fukushima-with-robots.html
[94] YOSHIYUKI SATO, et al., "Radiochemical analysis of rubble collected from around and inside reactor buildings at Units 1 to 4 in Fukushima Daiichi Nuclear Power Station", Hot Laboratories and Remote Handling Working Group (Proc. 54th Ann. Mtg Mito, 2017) JAEA, Tokyo (2017).
[95] KOMA, Y., et al., Radiochemical analysis of rubble and trees collected from Fukushima Daiichi Nuclear Power Station. J. Nucl. Sci. Tech. 51, (2014) 1032–1043.
[96] URLAND, C S., TMI-2 Postaccident Data Acquisition and Analysis Experience, Interim Report, USDOE, Washington, DC, 1992.
[97] AKERS, D.W., ROYBAL, G.S., Examination of Concrete Samples from the TMI-2 Reactor Building Basement, GEND-INF-081, EG&G Idaho, Idaho Falls, ID (1987).
[98] MEINKRANTZ, D.M., et al., First Results of Sump Samples Analyses — Entry 10, GEND-INF-011, EG&G Idaho, Idaho Falls, ID (1981).
[99] MCISAAC, C.V., et al., Results of Analyses Performed on Concrete Cores Removed from Floors and D-Ring Walls of the TMI-2 Reactor Building, GEND-054, EG&G Idaho, Idaho Falls, ID (1984).
[100] MCISAAC, C.V., KEEFER, D.G., TMI-2 Reactor Building Source Term Measurements: Surfaces and Basement Water and Sediment, GEND-042, EG&G Idaho, Idaho Falls, ID (1984).
[101] RUSSELL, M.L., TMI-2 Core Cavity Sides and Floor Examinations December 1985 and January 1986, GEND-074, EG&G Idaho, Idaho Falls, ID (1987).
[102] RUSSELL, M.L., et al., TMI-2 Accident Evaluation Programme Sample Acquisition and Examination Plan for FY87 and Beyond, EGG-TMI-7521, EG&G Idaho, Idaho Falls, ID (1987).
[103] BARATTA, A.J., GRICAR, B.G., JESTER, W.A., The Citizen Radiation Monitoring Programme for the TMI Area, GEND-008, Pennsylvania State University, University Park, PA (1981).
[104] KURINY, V.D., IVANOV, Y.A., KASHPAROV, V.A., et al., Particle associated Chernobyl fall-out in the local and intermediate zones. Ann. Nucl. Energy 20 (1993) 415–420.
[105] KASHPAROV, V.A, OUGHTON, D.H., ZVARICH, S.I., et al., Kinetics of fuel particles weathering and 90Sr mobility in the Chernobyl 30-km exclusion zone, Health Phys. 76 (1999) 251–259.
[106] KASHPAROV, V.A., et al., Kinetics of dissolution of Chernobyl fuel particles in soil in natural conditions. J. Environ. Radioact. 72 (2004) 335–353.
[107] BOBOVNIKOVA, Ts.I., VIRCHENKO, E.P., KONOPLEV, A.V., et al., Chemical forms of the long-living radionuclides and their transformation in the Chernobyl accident zone, Pochvovedeniye (Sov. Soil Sci.) 10 (1990) 20–25.
[108] KASHPAROV, V.A., LUNDIN, S.M., S.I. ZVARICH, Territory contamination with the radionuclides representing the fuel component of Chernobyl fallout, Sci. Total Environ. 317 (2003) 105–119.
[109] KASHPAROV, V.A., LUNDIN, S.M., KHOMUTININ, Y.V., et al., Soil contamination with 90Sr in the near zone of the Chernobyl accident, J. Environ. Radioactiv. 56 (2001) 285–298.
[110] ISTC SHELTER, Analysis of the Current Safety of the Shelter Object and Forecast Assessment of the Development of the Situation: Report/ISTC 'Shelter' NAS Ukraine, Arch. No. 3601, Chernobyl (1996).
[111] IVANOV, Y.A., KASHPAROV, V.A., Long-term dynamics of the radioecological situation in terrestrial Chernobyl exclusion zone, Environ. Sci. Pollut. Res. 10 (2003) 13–20.

[112] ALL RUSSIAN DESIGN AND RESEARCH INSTITUTE OF INDUSTRIAL TECHNOLOGY, Survey and Inventory of Section 2.1 PVLRO 'Red Forest' 1992, Research Report/VNIPIPT, Moscow, 1992.
[113] STATE ENTERPRISE SCIENTIFIC AND TECHNICAL CENTER OF DECONTAMINATION AND COMPLEX OF RADIOACTIVE WASTE MANAGEMENT, SUBSTANCES AND IONIZING RADIATION SOURCES, Research and Inventory of Section 5.1 PVLRO 'Neftebaza', STC KORO, Zhovti Vody (1994).
[114] ANTROPOV, V.M., et al. Review and Analysis of Solid Long-Lived and High Level Radioactive Waste Arising at the Chernobyl NPP and the Restricted Zone, EUR 19897, European Commission, Luxembourg (2001).
[115] BROWN, T.D., BILLON, F., Characterisation of Radioactive Waste Located at 'Shelter' Industrial Site, EUR 19844, European Commission, Luxembourg (2001).
[116] MINISTRY OF THE ENVIRONMENT, Guidelines for the Contamination Survey Method, Fact Sheet (2012), http://www.env.go.jp/press/files/jp/18929.pdf
[117] INTERNATIONAL ATOMIC ENERGY AGENCY, Cleanup and Decommissioning of a Nuclear Reactor After a Severe Accident, Technical Reports Series No. 346, IAEA, Vienna (1992).
[118] INTERNATIONAL ATOMIC ENERGY AGENCY, Planning for Cleanup of Large Areas Contaminated as a Result of a Nuclear Accident, Technical Reports Series No. 327, IAEA, Vienna (1991).
[119] INTERNATIONAL ATOMIC ENERGY AGENCY, Airborne Gamma Ray Spectrometer Surveying, Technical Reports Series No. 323, IAEA, Vienna (1991).
[120] INTERNATIONAL ATOMIC ENERGY AGENCY, Management of Severely Damaged Nuclear Fuel and Related Waste, Technical Reports Series No. 321, IAEA, Vienna (1990).
[121] INTERNATIONAL ATOMIC ENERGY AGENCY, Cleanup of Large Areas Contaminated as a Result of a Nuclear Accident, Technical Reports Series No. 300, IAEA, Vienna (1989).
[122] SELLAFIELD LTD, NUCLEAR DECOMMISSIONING AUTHORITY, Demolition Starts on Windscale chimney, Press Release (2019),
https://www.gov.uk/government/news/demolition-starts-on-windscale-chimney
[123] INTERNATIONAL ATOMIC ENERGY AGENCY, Retrieval and Conditioning of Solid Radioactive Waste from Old Facilities, Technical Reports Series No. 457, IAEA, Vienna (2007).
[124] INTERNATIONAL ATOMIC ENERGY AGENCY, Retrieval of Fluidizable Radioactive Wastes from Storage Facilities, IAEA-TECDOC-1518, IAEA, Vienna (2006).
[125] M.I. OJOVAN, Handbook of Advanced Radioactive Waste Conditioning Technologies, Woodhead Publishing Series in Energy No 12, Woodhead Publishing, Cambridge (2011).
[126] INTERNATIONAL ATOMIC ENERGY AGENCY, Categorizing Operational Radioactive Wastes, IAEA-TECDOC-1538, IAEA, Vienna (2007).
[127] FRENCH ALTERNATIVE ENERGIES AND ATOMIC ENERGY COMMISSION, Retour sur la Journée 'Fukushima, 4 ans après', News article (2015),
https://prositon.cea.fr/drf/prositon/Pages/Actualites/2015/actualit%C3%A9-20150519.aspx
[128] INTERNATIONAL ATOMIC ENERGY AGENCY, Technologies for In-situ Immobilization and Isolation of Radioactive Wastes at Disposal and Contaminated Sites, IAEA-TECDOC-972, IAEA, Vienna (1997).
[129] INTERNATIONAL ATOMIC ENERGY AGENCY, Treatment of Low- and Intermediate-Level Solid Radioactive Wastes, Technical Reports Series No. 223, IAEA, Vienna (1983).
[130] INTERNATIONAL ATOMIC ENERGY AGENCY, Treatment of Spent Ion-Exchange Resins for Storage and Disposal, Technical Reports Series No. 254, IAEA, Vienna (1985).
[131] INTERNATIONAL ATOMIC ENERGY AGENCY, Treatment of Alpha Bearing Wastes, Technical Reports Series No. 287, IAEA, Vienna (1988).
[132] INTERNATIONAL ATOMIC ENERGY AGENCY, Status of Technology for Volume Reduction and Treatment of Low and Intermediate Level Solid Radioactive Waste, Technical Reports Series No. 360, IAEA, Vienna (1994).
[133] INTERNATIONAL ATOMIC ENERGY AGENCY, Management of Waste Containing Tritium and Carbon-14, Technical Reports Series No. 421, IAEA, Vienna (2004).
[134] INTERNATIONAL ATOMIC ENERGY AGENCY, Management of Problematic Waste and Material Generated During the Decommissioning of Nuclear Facilities, Technical Reports Series No. 441, IAEA, Vienna (2006).
[135] INTERNATIONAL ATOMIC ENERGY AGENCY, Innovative Waste Treatment and Conditioning Technologies at NPPs, IAEA-TECDOC-1504, IAEA, Vienna (2006).
[136] INTERNATIONAL ATOMIC ENERGY AGENCY, Application of Thermal Technologies for Processing of Radioactive Waste, IAEA-TECDOC-1527, IAEA, Vienna (2006).
[137] INTERNATIONAL ATOMIC ENERGY AGENCY, Treatment and Conditioning of Radioactive Solid Wastes, IAEA-TECDOC-655, IAEA, Vienna (1992).

[138] INTERNATIONAL ATOMIC ENERGY AGENCY, The Volume Reduction of Low-activity Solid Wastes, Technical Reports Series No. 106, IAEA, Vienna (1970).
[139] MINISTRY OF THE ENVIRONMENT, Decontamination Report: A Compilation of Experiences to Date on Decontamination for the Living Environment Conducted by the Ministry of the Environment, FY2014, Ministry of the Environment, Tokyo, 2015.
[140] INTERNATIONAL ATOMIC ENERGY AGENCY, Treatment of Liquid Effluent from Uranium Mines and Mills, IAEA-TECDOC-1419, IAEA, Vienna (2004).
[141] INTERNATIONAL ATOMIC ENERGY AGENCY, Application of Membrane Technologies for Liquid Radioactive Waste Processing, Technical Reports Series No. 431, IAEA, Vienna (2004).
[142] INTERNATIONAL ATOMIC ENERGY AGENCY, Combined Methods for Liquid Radioactive Waste Treatment, IAEA-TECDOC-1336, IAEA, Vienna (2003).
[143] INTERNATIONAL ATOMIC ENERGY AGENCY, Treatment and Conditioning of Radioactive Organic Liquids, IAEA-TECDOC-656, IAEA, Vienna (1992).
[144] INTERNATIONAL ATOMIC ENERGY AGENCY, Treatment of Low- and Intermediate-level Liquid Radioactive Wastes, Technical Reports Series No. 236, IAEA, Vienna (1984).
[145] INTERNATIONAL ATOMIC ENERGY AGENCY, Chemical Precipitation Processes for the Treatment of Aqueous Radioactive Waste, Technical Reports Series No. 337, IAEA, Vienna (1992).
[146] INTERNATIONAL ATOMIC ENERGY AGENCY, Advances in Technologies for the Treatment of Low and Intermediate Level Radioactive Liquid Wastes, Technical Reports Series No. 370, IAEA, Vienna (1994).
[147] INTERNATIONAL ATOMIC ENERGY AGENCY, Application of Ion Exchange Processes for the Treatment of Radioactive Waste and Management of Spent Ion Exchangers, Technical Reports Series No. 408, IAEA, Vienna (2002).
[148] INTERNATIONAL ATOMIC ENERGY AGENCY, Handling and Treatment of Radioactive Aqueous Wastes, IAEA-TECDOC-654, IAEA, Vienna (1992).
[149] INTERNATIONAL ATOMIC ENERGY AGENCY, Waste Treatment and Immobilization Technologies Involving Inorganic Sorbents, IAEA-TECDOC-947, IAEA, Vienna (1997).
[150] TRIPLETT, M., Caesium Removal and Storage — Update on Fukushima Daiichi Status, Briefing for Hanford Advisory Board Tank Waste Committee (2015), http://www.hanford.gov/files.cfm/Attachment_6_Cs_Presentation_PNNL.pdf
[151] MINISTRY OF ECONOMY, TRADE AND INDUSTRY, The Outline of the Handling of ALPS Treated Water at Fukushima Daiichi NPS (FDNPS), Presentation (2019),
https://www.meti.go.jp/english/earthquake/nuclear/decommissioning/pdf/20191121_current_status.pdf
[152] NUCLEAR REGULATORY COMMISSION, Knowledge Management Library for the Three Mile Island Unit 2 Accident of 1979 (2016),
https://tmi2kml.inl.gov/HTML/Page1.html
[153] OAK RIDGE NATIONAL LABORATORY, Evaluation of the Submerged Demineralizer System (SDS) Flowsheet for Decontamination of High-Activity-Level Waste at the Three Mile Island Unit 2 Nuclear Power Station, ORNL/TM-7448, Oak Ridge National Laboratory, Oak Ridge, TN (1980).
[154] NATIONAL INSTITUTE FOR ENVIRONMENTAL STUDIES, Appropriate Treatment and Disposal of Disaster Waste and Waste Contaminated by Radioactive Substance (2022),
http://www.nies.go.jp/shinsai/1-1-e.html
[155] MIYAMOTO, Y., "R&D on the Radioactive Waste Treatment and Disposal", International Research Institute for Nuclear Decommissioning (Proc. Ann. Symp. Tokyo 2014), IRID, Tokyo (2014),
http://irid.or.jp/wp-content/uploads/2014/07/Sympo_Miyamoto_E.pdf
[156] MCCONNELL, J.W., ROGERS, R.D., "Results of field testing of waste forms using lysimeters", Waste Management '90 (Proc. Int. Symp. Tucson 1990) WM Symposia, Tucson, AZ, (1990).
[157] BRYAN, G.H., SIEMENS, D.H., "Development and Demonstration of a Process for Vitrification of TMI Zeolite", Transactions of the American Nuclear Society (ANS Winter Mtg, San Francisco, 1981) American Nuclear Society, La Grange Park, IL (1981).
[158] BARNER, J.O., DANIEL, J.L., MARSHALL, R.K., Zeolite Vitrification Demonstration Programme: Characterization of Radioactive Vitrified Zeolite Materials, GEND-INF-043, Pacific Northwest National Laboratory, Richland, WA, 1984.
[159] INTERNATIONAL ATOMIC ENERGY AGENCY, Handling and Processing Radioactive Waste from Nuclear Applications, Technical Reports Series No. 402, IAEA, Vienna (2001).

[160] INTERNATIONAL ATOMIC ENERGY AGENCY, Containers for Packaging of Solid Low and Intermediate Level Radioactive Wastes, Technical Reports Series No. 355, IAEA, Vienna (1993).
[161] INTERNATIONAL ATOMIC ENERGY AGENCY, Regulations for the Safe Transport of Radioactive Material, IAEA Safety Standards Series No. SSR-6 (Rev.1), IAEA, Vienna (2018).
[162] UNITED KINGDOM GOVERNMENT, NATIONAL WASTE PROGRAMME, Container Signposting Resource, Fact Sheet (2018),
https://www.gov.uk/government/publications/nwp-container-signposting-resource
[163] MINISTRY OF THE ENVIRONMENT, Guidelines for Designated Waste, Fact Sheet (2013),
http://www.env.go.jp/en/focus/docs/files/20140725-87-3.pdf
[164] UNITED STATES DEPARTMENT OF ENERGY, Accident Investigation Report: Phase 2 Radiological Release Event at the Waste Isolation Pilot Plant, 14 February 2014, Office of Environmental Management, Washington, DC (2015).
[165] WASTE ISOLATION PILOT PLANT, What Happened at WIPP in February 2014,
https://wipp.energy.gov/wipprecovery-accident-desc.asp
[166] MINISTRY OF FOREIGN AFFAIRS OF JAPAN, Events and Highlights on the Progress Related to Recovery Operations at Fukushima Daiichi Nuclear Power Station (2015),
https://www.mofa.go.jp/dns/inec/page22e_000222.html
[167] INTERNATIONAL ATOMIC ENERGY AGENCY, Handbook on the Storage of Radioactive Waste, IAEA Nuclear Energy Series, IAEA Vienna (in preparation).
[168] ROMANO, S. WELLING, S, BELL, S., "Environmentally Sound Disposal of Radioactive Materials at a RCRA Hazardous Waste Disposal Facility", Waste Management '03, (Proc. Int. Symp. Tucson 2003) WM Symposia, Tucson, AZ, (2003).
[169] INTERNATIONAL ATOMIC ENERGY AGENCY, The Radiological Accident in Goiânia, IAEA, Vienna (1988).
[170] DEPARTMENT OF EDUCATION, SCIENCE AND TRAINING, Rehabilitation of Former Nuclear Test Sites at Emu and Maralinga (Australia) 2003, Maralinga Rehabilitation Technical Advisory Committee, Commonwealth of Australia, Canberra (2002).
[171] MERRIL, E;GESELL, T.F., Environmental Radioactivity: From Natural, Industrial and Military Sources, Academic Press, London (1997) p. 429.
[172] LLERANDI, C.S., "Remediation after the Palomares accident: Scientific and social aspects", IAEA Report on Decommissioning and Remediation after a Nuclear Accident, Action Plan on Nuclear Safety Series , IAEA, Vienna (2013).
[173] EUROPEAN ATOMIC ENERGY COMMUNITY, FOOD AND AGRICULTURE ORGANIZATION OF THE UNITED NATIONS, INTERNATIONAL ATOMIC ENERGY AGENCY, INTERNATIONAL LABOUR ORGANIZATION, INTERNATIONAL MARITIME ORGANIZATION, OECD NUCLEAR ENERGY AGENCY, PAN AMERICAN HEALTH ORGANIZATION, UNITED NATIONS ENVIRONMENT PROGRAMME, WORLD HEALTH ORGANIZATION, Fundamental Safety Principles, IAEA Safety Standards Series No. SF-1, IAEA, Vienna (2006).
[174] THE NATIONAL ARCHIVES, Windscale Pile Incident October 1957: Report of Work Carried Out by the R&D Windscale Branch from 10 October 1957 to 5 November 1957, Ref. AB 7/6435, The National Archives, Kew (1957).
[175] UNITED KINGDOM ATOMIC ENERGY OFFICE, Accident at Windscale No. 1 Pile on 10th October, 1957, Nature 180 (1957) 1043.
[176] WAKEFORD, R., A double diamond anniversary — Kyshtym and Windscale: the nuclear accidents of 1957, J. Radiol. Prot. 37 (2017) E7.
[177] PENNEY, W., et al., Report on the accident at Windscale No. 1 Pile on 10 October 1957, J. Radiol. Prot., 37 (2017) 780–796.
[178] STEWART, N. G., CROOKS, R.N., Long-range travel of the radioactive cloud from the accident at Windscale, Nature 182 (1958) 627–8.
[179] GARLAND, J.A., WAKEFORD, R., Atmospheric emissions from the Windscale accident of October 1957, Atmos. Environ. 41 (2007) 3904–20.
[180] RUTHERFORD, T.N., The Windscale Piles: Situation April 1961, May 1961, Ref. AB62/71, The National Archives, London (1961).
[181] UNITED KINGDOM HEALTH AND SAFETY EXECUTIVE, Guidance for Inspectors on the Management of Radioactive Materials and Radioactive Waste on Nuclear Licensed Sites, UK HSE, Nuclear Safety Directorate, London (2001).

[182] ELECTRIC POWER RESEARCH INSTITUTE, The Cleanup of Three Mile Island Unit 2 — A Technical History: 1979–1990, Report NP-6931, EPRI, Palo Alto, CA (1990).
[183] ELECTRIC POWER RESEARCH INSTITUTE, TMI-2 Waste Management Experience, Report TR-100640, EPRI, Palo Alto, CA (1992).
[184] NUCNET, NucNet's Updated 'Chernobyl Fact File' Now Online, News Article (2016), https://www.nucnet.org/news/nucnet-s-updated-chernobyl-fact-file-now-online
[185] KISELEV, A.N., CHERCHEROV, K.P., Model of the destruction of the reactor in the No. 4 unit of the Chernobyl NPP, Atomic Energy 91 (2001) 967–975.
[186] OSKOLKOV, B.Y., et al., Radioactive waste management in the Chernobyl exclusion zone: 25 years since the Chernobyl NPP accident, Health Phys. 101 (2011) 431–441.
[187] KASHPARVO, V.O., "The formation and dynamics of radioactive contamination of the environment during the accident at Chernobyl NPP and in the post-accident period", Chernobyl: The Exclusion Zone: Digest of Research Papers (Bariakhtar, V., Ed.) Naukova Dumka, Kiev (2001) pp. 11–46.
[188] SAMOILENKO, Y.N., GOLUBEV, V.V., "Decontamination of the 'special zone'", Chernobyl 88. Reports of the 1st All-Union Scientific-Technical Meeting on the Outcomes of Eliminating Aftermaths of the Accident at Chernobyl NPP, GNTU, PO 'Kombinat', Ministry of Atomic Energy of the USSR (1989) pp. 205–225.
[189] ILYIN, L.N., "Chemical troops", Chernobyl: Disaster. Deed, Lessons and Conclusions (Dyachenko, A.A., Ed.), Inter-Vesy, Moscow (1996) pp. 523–552.
[190] ANTROPOV, V.M., KUMSHAEV, S.B., SKVORTSOV, V.V., KHABRIKA, A.I., Clarification of data on radioactive waste placed in storage facilities of the Chernobyl Exclusion Zone, Bulletin of the ecological status of the exclusion zone and the zone of unconditional (mandatory) resettlement, GAZO 2 (2004) 24.
[191] ANTROPOV, V.M., MELNICHENKO, V.P., TRETYAK, O.G., KHABRIKA, A.I., Analysis of the Condition of Radioactive Waste in the Trenches of PVLRO 'Neftebaza' (Antropov, et al., Ed.) KIM Publishing House, Kyiv (2012).
[192] EICHHORN, H. Industrial Complex for Solid Radwaste Management (ICSRM) at Chernobyl NPP functionality of the facilities. Factors of success, Atw. Internationale Zeitschrift fuer Kernenergie 57 (2011) 105–107.
[193] INTERNATIONAL ATOMIC ENERGY AGENCY, The Fukushima Daiichi Accident, IAEA, Vienna (2015).
[194] INTERNATIONAL ATOMIC ENERGY AGENCY, International Peer Review Mission on Mid-and-Long-Term Roadmap Towards the Decommissioning of TEPCO's Fukushima Daiichi Nuclear Power Station Units 1–4: Fourth Mission: Preliminary Summary Report to the Government of Japan, IAEA, Vienna (2018).
[195] BATORSHIN, G., MOKROV, Y.G., "Experience in Eliminating the Consequences of the 1957 Accident at the Mayak Production Association", paper presented at Int. Expert Mtg on Decommissioning and Remediation after a Nuclear Accident, IAEA, Vienna 2013.
[196] MERKUSHKIN, A.O., "Karachay lake is the storage of the radioactive wastes under open sky", presented at Int. Youth Nucl. Congress 2000: Youth, Future, Nuclear Proceedings and Multimedia Presentation, YDRNS Information Technology Services, Russian Federation, 2001.
[197] WORLD NUCLEAR ASSOCIATION, Chernobyl Accident 1986, Fact Sheet (2020), https://www.world-nuclear.org/information-library/safety-and-security/safety-of-plants/chernobyl-accident.aspx
[198] IRANZO, E., ESPONOSA, A., IRANZO, C.E., "Evaluation of remedial actions taken in an agricultural area contaminated by transuranides", Radioecology: The Impact of Nuclear Origin Accidents on Environment (Proc. IV Int. Symp. Cadarache, 1988), CEA Centre d'Etudes Nucleaires, Cadarache (1988).
[199] MOLINA, G., "Lessons learned during the recovery operations in the Ciudad Juarez accident", Recovery Operations in the Event of A Nuclear Accident or Radiological Emergency, Proceedings Series, IAEA, Vienna (1990).
[200] LEHTO, J., PAAJANEN, A.A., Review of cleanup of large radioactive-contaminated areas, Cleanup of Large Radioactive Contaminated Areas and Disposal of Generated Waste (Lehto, J., Ed.), vol. 1994/567, TemaNord (1994) pp. 3–21.
[201] AMARAL, E., "Remediation following the Goiânia accident", paper presented at Int. Expert Mtg on Decommissioning and Remediation after a Nuclear Accident, IAEA, Vienna 2013.
[202] PONTEDEIRO, E.M., HEILBRON, P.F., PEREZ-GERRERO, J. et al., Reassessment of the Goiânia radioactive waste repository in Brazil using HYDRUS-1D, J. Hydrol. Hydromech. 66 (2018).
[203] ARNAL, J.M., et al., Management of Radioactive Ashes after a ^{137}Cs Source Fusion Incident (2004), http://irpa11.irpa.net/pdfs/7e5.pdf
[204] NUCLEAR DECOMMISSIONING AUTHORITY, Nuclear Decommissioning Authority: Strategy (2016), https://www.gov.uk/government/consultations/nuclear-decommissioning-authority-draft-strategy
[205] NUCLEAR DECOMMISSIONING AUTHORITY, Radioactive Wastes in the UK: A Summary of the 2016 Inventory, Nuclear Decommissioning Authority, Moor Row (2017).

[206] NUCLEAR DECOMMISSIONING AUTHORITY, Land Quality Management. Preferred Option (Gate B), Ref: SMS/TS/A3-LQM/001/B, Nuclear Decommissioning Authority, Moor Row, 2011.

[207] NUCLEAR DECOMMISSIONING AUTHORITY, UK Strategy for the Management of Solid Low Level Radioactive Waste from the Nuclear Industry, Report (2010), https://assets.publishing.service.gov.uk/government/uploads/system/uploads/attachment_data/file/457083/UK_Strategy_for_the_Management_of_Solid_Low_Level_Radioactive_Waste_from_the_Nuclear_Industry_August_2010.pdf

[208] DEPARTMENT OF ENERGY AND CLIMATE CHANGE, UK Strategy for the Management of Solid Low Level Waste from the Nuclear Industry, URN 15D/472, Department of Energy and Climate Change, Kew, 2016.

[209] NUCLEAR INDUSTRY SAFETY DIRECTORS FORUM, Clearance and Radiological Sentencing: Principles, Processes and Practices, A Nuclear Industry Guide, Guide (2017), https://www.nuclearinst.com/write/MediaUploads/SDF%20documents/CEWG/Clearance_and_Exemption_GPG_2.01.pdf

[210] HAWKINS, A.R., Hanford Regulatory Experience Regulation at Hanford — A Case Study, US Department of Energy, US Department of Energy, DOE-0333-FPNA, Richland, WA (2007).

[211] WILLIAMS, G. Remediation of Contaminated Lands: the Maralinga lessons, Australian Radiation Protection and Nuclear Safety Agency (ARPANSA), Presentation (2013), https://www-pub.iaea.org/iaeameetings/IEM4/31Jan/Williams.pdf

[212] SOLENTE, N., "VLLW disposal and management of large volume of slightly contaminated materials: The French experience", paper presented at IAEA Technical Meeting on the Disposal of Large Volume of Radioactive Waste, IAEA, Vienna (2013).

ABBREVIATIONS

ALPS	advanced liquid processing system
ASKRO	Automated Radiation Monitoring System
CFR	Code of Federal Regulations
CIRES	Industrial Centre for Collection, Storage and Disposal
CODIRPA	Comité Directeur pour la gestion de la phase post-accidentelle d'un accident nucléaire ou d'une situation radiologique
CZT	cadmium telluride implanted zinc
DA	destructive assay
DAW	dry activated waste
DOE	US Department of Energy
DOT	US Department of Transportation
DQO	data quality objective
ENSDF	Engineered Near Surface Disposal Facility
EP	electrical precipitator
EPRI	Electric Power Research Institute
ERDF	Environmental Restoration Disposal Facility
EW	exempt waste
FCM	fuel containing material
GDR	gamma dose rate
GEND	General Public Utilities, Electric Power Research Institute, US Nuclear Regulatory Commission and the US Department of Energy
GPS	global positioning satellite
GPU	General Public Utilities
GTCC	greater than class C
HIC	high integrity container
HLW	high level waste
ICSRM	Industrial Complex for Solid Radwaste Management
ILW	intermediate level waste
INES	International Nuclear and Radiological Event Scale
IRID	International Research Institute for Nuclear Decommissioning
ISF	interim storage facility
ISOCS	In Situ Object Counting System
ISRWR	Installation for Solid Radioactive Waste Recovery
JAEA	Japan Atomic Energy Agency
JNES	Japan Nuclear Energy Safety Organization
LILW	low and intermediate level waste
LLW	low level waste
LRTP	Liquid Radwaste Treatment Plant
MOE	Ministry of Environment, Japan
NDA	non-destructive assay
NISA	Nuclear and Industrial Safety Agency, Japan
NPP	Nuclear Power Plant
NRA	Nuclear Regulation Authority, Japan
NRC	US Nuclear Regulatory Commission
NSC	New Safe Confinement
NSF	near surface facility
OG	Operations Group

PPE	personal protective equipment
QA	quality assurance
QC	quality control
R&D	research and development
RBMK	high-power channel-type reactor
RMS	requirements management system
RW	radioactive waste
RWDS	radioactive waste disposal site
RWTSP	radioactive waste temporary storage place
SAFSTOR	safe enclosure
SARRY	Simplified Active Water Retrieve and Recovery System
SAUEZM	State Agency of Ukraine on Exclusion Zone Management
SDA	special decontamination area
SDS	submerged demineralizer system
SIP	Shelter Implementation Plan
SRWSF	Solid Radioactive Waste Storage Facility
SRWTP	Solid Radioactive Waste Treatment Plant
SSC	structure, system and component
SSE	special state enterprise
TEPCO	Tokyo Electric Power Company
TMI	Three Mile Island
TRU	transuranic
UAV	unmanned aerial vehicle
UKAEA	United Kingdom Atomic Energy Authority
USFCRFC	Ukrainian Society for Friendship and Cultural Relations with Foreign Countries
VLLW	very low level waste
VNIPIPT	All-Russia Development and Scientific Research Institute for Industrial Technology
VSLW	very short lived waste
WAC	waste acceptance criteria
WIPP	Waste Isolation Pilot Plant

CONTRIBUTORS TO DRAFTING AND REVIEW

Ashida, T.	Japan Atomic Energy Agency, Japan
Barboa, J.	United States of America
Beckman, D.	Beckman and Associates, United States of America
Bondarkov, M.	State Agency of Ukraine on Exclusion Zone Management, Ukraine
Brennecke, P.	Germany
Chapman, N.	Ireland
Clifford, J.	Sellafield Ltd, UK
Drace, Z.	International Atomic Energy Agency
Dragolici, F.	International Atomic Energy Agency
Durham, L.A.	Environmental Science Division, Argonne National Laboratory, USA
Dutzer, M.	ANDRA, France
Garamszeghy, M.	Nuclear Waste Management Organization, Canada
Green, T.	Nuvia Limited, UK
Grogan, H.	Cascade Scientific, Inc., United States of America
Inoue, T.	Central Research Institute of Electric Power Industry, Japan
Kuchynskyi, V.	SSE Chornobyl NPP, Ukraine
Kumalo, Y.	International Atomic Energy Agency
Marra, J.	Savannah River National Laboratory, US Department of Energy, USA
Masaki, K.	Tokyo Electric Power Company Holdings
Mayer, S.J.	International Atomic Energy Agency
Miller, B.	AMEC, UK
Ojovan, M.I.	International Atomic Energy Agency
Ormai, P.	International Atomic Energy Agency
Phathanapirom, U.	International Atomic Energy Agency
Pillette-Cousin, L.	ANDRA, France
Poisson, R.	ANDRA, France
Prevost, T.	AREVA, France
Robbins, R.A.	International Atomic Energy Agency
Romano, S.	American Ecology Group, USA

Samanta, S.K.	International Atomic Energy Agency
Schultheisz, D.	US Environmental Protection Agency, United States of America
Shimba-Yamada, M.	International Atomic Energy Agency
Sizov, A.	Institute for Safety Problems of NPP, National Academy of Sciences, Ukraine
Skrypov, M.	Special State Enterprise Chornobyl NPP, Ukraine
Solente, N.	ANDRA, France
Tokarevsky, V.	Institute for Chornobyl Problems, Ukraine
Tsurikov, N.	Calytrix Consulting Pty Ltd, Australia
Walsch, C.	Sellafield Ltd
Yagi, M.	International Atomic Energy Agency
Yamada, K.	National Institute for Environmental Studies, Japan

Technical and Consultants Meetings

Vienna, Austria: 14–17 May 2013, 18–21 November 2013,
25–28 November 2013, 23–27 June 2014, 9–13 March 2015,
24–28 August 2015, 18–21 July 2017, 27–31 May 2019,
Washington, DC: 2–6 March 2015

Structure of the IAEA Nuclear Energy Series*

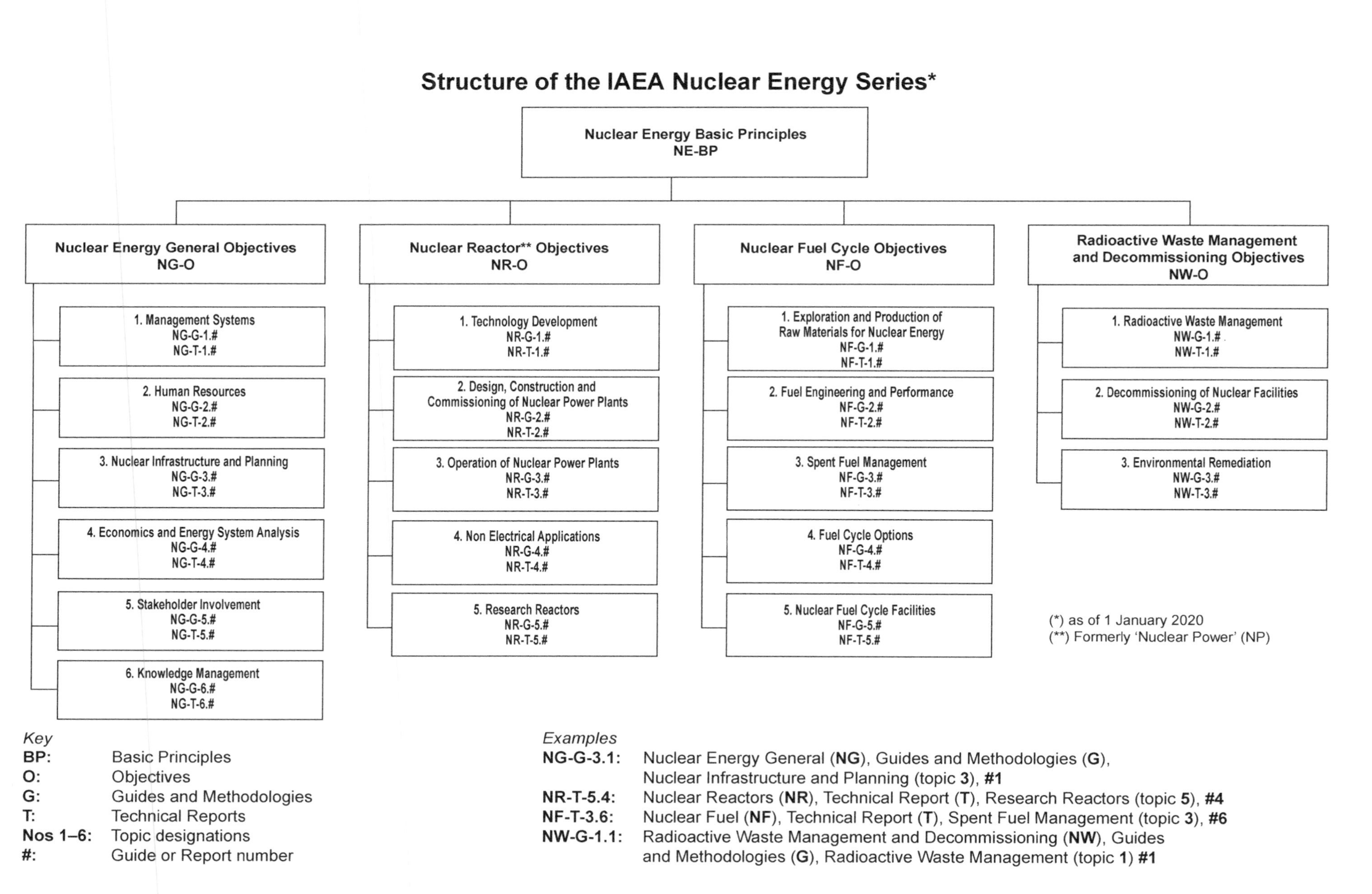

(*) as of 1 January 2020
(**) Formerly 'Nuclear Power' (NP)

Key

BP: Basic Principles
O: Objectives
G: Guides and Methodologies
T: Technical Reports
Nos 1–6: Topic designations
#: Guide or Report number

Examples

NG-G-3.1: Nuclear Energy General (**NG**), Guides and Methodologies (**G**), Nuclear Infrastructure and Planning (topic **3**), **#1**
NR-T-5.4: Nuclear Reactors (**NR**), Technical Report (**T**), Research Reactors (topic **5**), **#4**
NF-T-3.6: Nuclear Fuel (**NF**), Technical Report (**T**), Spent Fuel Management (topic **3**), **#6**
NW-G-1.1: Radioactive Waste Management and Decommissioning (**NW**), Guides and Methodologies (**G**), Radioactive Waste Management (topic **1**) **#1**

No. 26

ORDERING LOCALLY

IAEA priced publications may be purchased from the sources listed below or from major local booksellers.

Orders for unpriced publications should be made directly to the IAEA. The contact details are given at the end of this list.

NORTH AMERICA

Bernan / Rowman & Littlefield

15250 NBN Way, Blue Ridge Summit, PA 17214, USA
Telephone: +1 800 462 6420 • Fax: +1 800 338 4550
Email: orders@rowman.com • Web site: www.rowman.com/bernan

REST OF WORLD

Please contact your preferred local supplier, or our lead distributor:

Eurospan Group

Gray's Inn House
127 Clerkenwell Road
London EC1R 5DB
United Kingdom

Trade orders and enquiries:

Telephone: +44 (0)176 760 4972 • Fax: +44 (0)176 760 1640
Email: eurospan@turpin-distribution.com

Individual orders:

www.eurospanbookstore.com/iaea

For further information:

Telephone: +44 (0)207 240 0856 • Fax: +44 (0)207 379 0609
Email: info@eurospangroup.com • Web site: www.eurospangroup.com

Orders for both priced and unpriced publications may be addressed directly to:

Marketing and Sales Unit
International Atomic Energy Agency
Vienna International Centre, PO Box 100, 1400 Vienna, Austria
Telephone: +43 1 2600 22529 or 22530 • Fax: +43 1 26007 22529
Email: sales.publications@iaea.org • Web site: www.iaea.org/publications

22-02768E-T